Grave Trippers II

Joel!
Happy Grave Tripping!!
Vincent & Robert
10/30/24

Grave Trippers II

REMEMBERING AND RESPECTING
AMERICA'S HEROES AND HEROINES

Robert and Vincent Gardino

Camino Books, Inc.
PHILADELPHIA

Copyright © 2024 by Robert Gardino and Vincent Gardino

All rights reserved.

Camino Books® and its logo are registered trademarks of Camino Books, Incorporated

No part of this book may be reproduced in any form or by any electronic or mechanical means including information storage and retrieval systems without permission in writing from the publisher, except by a reviewer who may quote brief passages in a review.

Manufactured in the United States of America

1 2 3 4   27 26 25 24

Library of Congress Cataloging-in-Publication Data

Names: Gardino, Robert, 1957– author. | Gardino, Vincent, 1953– author.
Title: Grave trippers II, Remembering and respecting America's heroes and heroines / Robert and Vincent Gardino.
Other titles: Remembering and respecting America's heroes and heroines
Description: Philadelphia : Camino Books, Inc., [2024]
Identifiers: LCCN 2024014547 (print) | LCCN 2024014548 (ebook) | ISBN 9781680980516 (trade paperback) | ISBN 9781680980523 (ebook)
Subjects: LCSH: Cemeteries—United States—Guidebooks. | Heroes—United States—Biography—Anecdotes. | Celebrities—United States—Biography—Anecdotes. | Celebrities—Tombs—United States—Guidebooks. | Historic sites—United States—Guidebooks.
Classification: LCC E159 .G2183 2024 (print) | LCC E159 (ebook) | DDC 363.7/50973—dc23/eng/20240329
LC record available at https://lccn.loc.gov/2024014547
LC ebook record available at https://lccn.loc.gov/2024014548

ISBN 978-1-68098-051-6
ISBN 978-1-68098-052-3 (ebook)

Interior design: P. M. Gordon Associates, Inc.

Cover design: Jerilyn DiCarlo

This book is available at a special discount on bulk purchases for educational, business, and promotional purposes. For information write:

Publisher
Camino Books, Inc.
P.O. Box 59026
Philadelphia, PA 19102
www.caminobooks.com

*To our devoted physicians,
who help keep us above ground*

Dr. Sun-Hi Lee

Dr. Stanley Fahn

Dr. Kimberly Kwei

Dr. Craig Smith

Dr. Brett Youngerman

Dr. Wayne Whitmore

Contents

FOREWORD xi

INTRODUCTION 1

CHAPTER ONE

Congressional Medal of Honor Winners 3

SYLVESTER ANTOLAK   4
EDWARD CARTER   6
AUDIE MURPHY   8
VINCENT ROBERT CAPODANNO   11
THEODORE ROOSEVELT   13
RICHARD BYRD   19
ALVIN YORK   22

CHAPTER TWO

U.S. Presidents 27

CHESTER ALLEN ARTHUR   28
HERBERT HOOVER   31

CHAPTER THREE

U.S. Government Officials 35

WILLIAM SEWARD   36
FIORELLO LA GUARDIA   38

ALEXANDER HAMILTON   41
CLARE BOOTH LUCE   45
EDWIN STANTON   48
RAYMOND DONOVAN   53

CHAPTER FOUR
Astronauts/Pilots   57

SALLY RIDE   58
AMELIA EARHART   60
VIRGIL "GUS" GRISSOM   64
ED WHITE   67
NEIL ARMSTRONG   71
FRANCIS GARY POWERS   74

CHAPTER FIVE
Sports   79

BABE RUTH   80
JIM THORPE   84
JOE DiMAGGIO   87
JACKIE ROBINSON   90
HANK AARON   97
JESSE OWENS   93
HANK AARON   97
MABEL FAIRBANKS   100
VINCE LOMBARDI   102
WILMA RUDOLPH   105

CHAPTER SIX
Entertainment   109

HARRY HOUDINI   110
JUDY HOLLIDAY   113
MARY TYLER MOORE   116
ANNA MAY WONG   120

CONTENTS ix

HATTIE McDANIEL 123
FRANK SINATRA 126
ED SULLIVAN 130
EDDIE RABBITT 133
FANNIE BRICE 135
PAUL NEWMAN 137
BOB HOPE 138
MARIAN ANDERSON 141
WILLIAM S. PALEY 143
MARIE TORRE 147

CHAPTER SEVEN
Clergy 157
ARCHBISHOP FULTON SHEEN 152
CARDINAL EDWARD EGAN 154
REVEREND NORMAN VINCENT PEALE 158

CHAPTER EIGHT
Judiciary 161
LEARNED HAND 162
SAMUEL SEABURY 165
OLIVER WENDELL HOLMES, JR. 168

AFTERWORD 173

ACKNOWLEDGMENTS 175

Foreword

LOUIS L. PICONE

The authors of this book, Robert and Vincent, and I have a lot in common: we enjoy history and a good Italian meal, but most relevant to this Foreword, we have each spent a *lot* of time with dead people. We have independently visited cemeteries throughout the country and have discovered there is a lot to be learned from these hallowed grounds. While writing the books *The President Is Dead!* and *Grant's Tomb*, my research brought me to the graves of every deceased president of the United States. Their final resting places are as unique as the men themselves. They reflect the presidents and how the public remembered them, but also offer insight into American history. The humble graves of George Washington, Thomas Jefferson, and James Madison at their family estates suggest their deep sense of "small-r" republican virtues, a belief that presidents should not be deified, but rather after their service, they should return to using the greatest title, American citizen. Through their simplicity, we see how they viewed the presidency as distinct from the royals of Europe.

The antebellum period ushered in a movement toward natural, rural cemeteries, some designed by the great architects of the day. Examples or rural repose such as Crown Hill Cemetery in Indianapolis (where President Benjamin Harrison is buried); Hollywood Cemetery in Richmond (where both Presidents John Tyler and James Monroe are buried); and Albany Rural Cemetery in New York (where President Chester Alan Arthur is buried) offered mourners natural beauty and spiritual serenity.

Union victory in the Civil War ushered in a period of economic prosperity known as the Gilded Age. America rivaled European empires for opulence, but sadly was also afflicted with what had previously been considered a foreign scourge, political assassinations. Massive public mourning and the popularity of neoclassical architecture resulted in the colossal tombs of Abraham Lincoln, James Garfield, and William McKinley. While he did not die in office, the tomb of Ulysses S. Grant, the largest in America to this day (as per the National Park Service) indicates his immense and universal popularity at the time of his death. While none of these men left any instructions as to their monumental resting places, those who did opted for simple graves in contrast to, for example, that of the emerging imperial presidency, notably Theodore Roosevelt's grave, more than likely planned by a family member.

Since World War II and the establishment of Presidential Libraries, our presidents' graves, such as Herbert Hoover's, have been more frequently located on the grounds of the research and archival institutions that, in effect, have become their monuments. It is interesting to ponder where history, and presidential graves, will go from here.

Needless to say, I was delighted when my good friends Robert and Vincent invited me to accompany them on a grave-tripping expedition. I believe I accepted before they finished the question, but as if extra incentive was required, we would be visiting Oak Hill Cemetery in Washington, DC, where Abraham Lincoln's son Willie had been temporarily interred. The somber crypt in a scenic corner of the cemetery rests at the foot of a precarious stone staircase. As we carefully descended, the Gardinos explained how President Lincoln had the coffin opened several times so he could look upon his beloved son's face. One can still sense the unbearable pain and anguish of Lincoln mourning his son, while simultaneously fighting a Civil War to save the Union and abolish slavery.

After several moments of contemplation and the obligatory photographs, we continued our tour. I followed them to the final resting places of such notables as Secretaries of State Dean Acheson (under the term of Harry Truman) and Madeleine Albright (Bill Clinton); *Washington Post* leaders during the Watergate scandal, Ben Bradlee and Katherine Graham; Abraham Lincoln's Secretary of War, Edwin

M. Stanton; and the woman at the center of the Peacoat Affair during Andrew Jackson's administration, Peggy Eaton. In true gravetripper fashion, we got lost more than once, but also stumbled upon a few unexpected surprises such as the grave of Surgeon General Joseph Barnes, who helped care for Lincoln in his final hours; Lincoln's private secretary, John Nicolay; and funeral director to the presidents Joseph Gawler. Our time at Oak Hill, or any cemetery for that matter, enabled us to commune with these historic figures in death as we no longer can, or ever could, in life.

At each stop, the Gardino brothers reeled off fascinating personal details, revealing historic context and enlightening anecdotes. Robert and Vincent have a passion for the complex tapestry of American history and its many facets such as politics, military issues, news media, sports, religion, and popular culture. The Gardinos also have a unique appreciation for the men and women, sung and unsung, who have sacrificed and struggled along America's endless journey toward a more perfect union, which is the theme of *Grave Trippers II*. Among the individuals they chose to highlight, I am personally thrilled to find three presidents: Theodore Roosevelt, Chester Alan Arthur, and Herbert Hoover. While Roosevelt's accomplishments on both the military and political battlefields are widely celebrated, less so are the consequential and courageous achievements of Arthur, by his passing the Pendleton Act, and Hoover's efforts to prevent starvation in Europe in both World Wars.

These are just a few of the dozens of deserving men and women featured in *Grave Trippers II*. Some are well-known, while others you may discover for the first time. But whether you read *Grave Trippers II* in the comfort of your home, or stuff it in your back pocket as you set off on your own cemetery adventure, you are sure to discover someone compelling to meet on every page, and something fascinating to learn at every grave.

Introduction

All the individuals covered in this second volume of Grave Trippers made positive contributions that we feel should not be forgotten. The dictionary describes a hero/heroine as an individual who is much admired for great or brave acts in the face of adversity. We feel the acts of heroism of those in this volume should continue to inspire future generations, and it is our privilege to share their stories.

Ask people what general characteristics make up a hero or a heroine, and there is little doubt topping the list would be courage. It certainly takes a great deal of courage to risk one's life to save someone, for example, from a burning building. Volunteering to go on a dangerous military mission certainly can be described as a courageous act. But courage is not the only characteristic that makes a man or woman a hero. Generosity and a determination to do the right thing can make an individual a hero. A person can be a hero for having shown deep perseverance in overcoming great obstacles or disabilities. There are numerous other qualities that can define someone as heroic, as you will see regarding the individuals covered in this book.

We did our best to highlight men and women from various walks in life: entertainers, presidents, judges, military personnel, clergy and sports figures. Some of the names will be immediately recognizable, while other names are more obscure. In fact, it was one of those obscure names in our first book, Margaret Corbin, that served as an inspiration for the heroic theme of this book. Briefly, Corbin was an American Revolutionary War heroine. After her husband was killed before her eyes during the Battle of Fort Washington, she took over the cannon her husband had been manning and fired on the advancing British soldiers. She was severely wounded and captured by the British, but due to her wounds was released. For her heroism the Continental Congress awarded her half of a soldier's monthly pay, in essence making Margaret Corbin the first woman to collect a military pension.

When our first book, *Grave Trippers: History at Our Feet*, came out we appreciated many of the positive reviews it received. However, every now and then we would meet someone who would comment, "How come you didn't cover so and so at one of the cemeteries you highlighted?" At the conclusion of our book presentations when we would take questions from the audience, inevitably someone would ask, "Will there be a *Grave Trippers II*?" The answer to the second question is obviously yes. Our response to the first question is that we cannot guarantee that we have covered everybody's favorite notable individual, but we tried to cover an eclectic collection of people from numerous professions. Some individuals will be controversial. We cannot emphasize enough that just because we consider an individual a hero or a heroine does not make them a saint. They were imperfect human beings just like the rest of us. But despite their individual frailties and weaknesses, these people made significant positive contributions in the lives of those who were impacted by their actions.

In our first book, we covered cemeteries and individual graves that were located just on the East Coast. Now in this volume, we cover graves not only on the East Coast but also on the West Coast and a few graves in between. Also, this book has a similar format to our previous volume. The only difference is that we dispense with a brief history of the cemetery or graveyard. Short biographies of all individuals profiled is the primary focus of this book. We then provide accurate directions to graves within the cemetery or graveyard so that they can be more easily found. We also feature some of our prized autographs. Finally, all of those profiled include fun facts—interesting tidbits that most people do not know.

We certainly hope that you find this second book on our hobby of visiting the graves of historical or notable persons interesting. If you find the stories of the individuals covered herein inspirational or if any of them reignites an interest in history, then we will feel that we have accomplished our mission. Enjoy!

CHAPTER ONE

Congressional Medal of Honor Winners

Graciously hearken to us as soldiers who call Thee that, armed with Thy power, we may advance from victory to victory, and crush the oppression and wickedness of our enemies, and establish Thy justice among men and nations.

Written at the request of General George S. Patton by his Chaplain, James Hugh O'Neil

Greater love hath no man than this, that a man lay down his life for his friends.

John 15:13

SYLVESTER ANTOLAK
U.S. Army Sergeant
Born: September 10, 1918
Died: May 24, 1944

Words, either written or spoken, can never adequately describe the true valor displayed by this Congressional Medal of Honor awardee. Valor is defined as great courage in the face of danger, especially in battle.

Sylvester Antolak was born on a farm in St. Clairsville, Ohio and was of Polish descent. Having grown up as a boy in the '30s during the Great Depression, he understood the meaning of hard work and the necessity of assisting his parents on their farm. At the age of 22, Antolak enlisted in the army in 1941, where he achieved the rank of sergeant and was assigned to lead the First Platoon of Company B, which was part of the Infantry Division. Sergeant Antolak and his men were sent in May 1944 to Anzio, Italy, which is a coastal area less than 10 miles south of Rome. The mission was to eliminate German defensive lines and pave the way for a clear path to Rome, which was at the time controlled by the Axis Powers.

Near the town of Cisterna, Antolak and his men were pinned down by German machine gun fire. Their cover was minimal, and to take out the Germans' position, between them was basically an open field of approximately 200 yards. Realizing that to stay put was not a viable option, and that such a decision might lead to his entire company being wiped out, Antolak took decisive action. In order to cause a distraction, Antolak, armed with his own machine gun rifle, began to run as fast as he possibly could in the opposite direction from where his company was pinned down. His running in a zigzag pattern and firing his submachine gun at the German nest position gained Antolak approximately 30 yards that must have been a wonder not only to his men but to the Germans as well. However, enemy fire finally found their target, and Antolak was hit. He was down on the ground, and his men were yards away, unable to tell whether their platoon leader was alive or not. But after what might have seemed an eternity, his men saw Antolak stand up and signal to them to charge with him toward the German position. Antolak once again resumed his run toward the enemy fire, and for a second time went down, this time with his right shoulder shattered by enemy fire.

Sylvester Antolak's Grave

Displaying what seemed to be superhuman strength, Antolak rose for a second time and continued along with his men, running as fast as he could toward the German position. When he was approximately 15 yards from their nest, Antolak fired into the German position, killing two Germans and thereby forcing the remaining stunned ten Germans to surrender. Antolak's own men then pleaded with him to get the medical attention he so desperately needed. But instead, he focused his attention on a second German defensive position approximately 100 yards away. Incredibly, as if he felt no pain at all from his multiple bullet wounds, he charged the second German position and got within 25 yards before German concentrated fire killed him. Antolak's men were enraged when they saw their platoon leader killed. The remaining members of Company B went on to finish their mission by charging the second German position, and took out or captured a total of 20 German soldiers that day.

Sergeant Sylvester Antolak was awarded the Congressional Medal of Honor posthumously. The medal was given to his mother, who was informed of her son's death on her Ohio farm.

FUN FACT

Among the soldiers led by Antolak, who witnessed his heroics, was fellow Congressional Medal of Honor winner Audie Murphy (who is profiled later).

CEMETERY

Buried at Sicily-Rome American Cemetery, Piazzale Kennedy 1, 00048 in Nettino, Lazio, Italy. Plot C, Row 12, Grave 13.

EDWARD CARTER

U.S. Army Sergeant
Born: May 26, 1916
Died: January 30, 1963

Edward A. Carter, Jr. was born in Los Angeles, California. He was the product of an African American father and an East Indian mother. He grew up mostly in India and later In Shanghai, China. He was fluent not only in English, but also in Hindi, German and Mandarin. Carter entered the American army in September of 1941.

Because he was an African American serving at a time when there was the ridiculous belief that black individuals could not be trained for combat duty, Carter was relegated to administrative duty, where he achieved the rank of staff sergeant. Following huge American losses incurred at the Battle of the Bulge, however, the necessity for additional men for combat duty gave Carter his chance to engage in the fighting. By volunteering for provisional duty, Carter had to relinquish his rank of staff sergeant and be demoted back to private.

In March of 1945, close to Speyer, Germany, Carter volunteered to lead other black servicemen riding in a tank that came under German bazooka and small fire attack that was close to a warehouse. Carter led his three-man unit toward the warehouse after spotting where the

Edward Carter's Grave

bazooka fire was coming from. From their covered positions, Carter and his men had to cross an open field to advance toward the German position. When he gave his three men the go-ahead to follow him toward the enemy defensive line, which was approximately 150 yards away, they came under intense firing, and one was instantly killed. Carter then ordered his remaining two men to return to their original cover position so that he might proceed alone. However, enemy fire killed a second man and seriously wounded the third as they returned, trying to get back to their cover. Carter proceeded toward the fire and got hit three times in his left arm. Then he was knocked off his feet as he was hit in his left leg. Carter once again was hit when he tried to take a drink of water from his canteen. The bullet went straight through his left hand. Despite his wounds, Carter continued advancing on the German position, crawling to within 30 yards of it. The enemy fire became so heavy that Carter had to remain behind an embankment for about two hours. At that time eight German soldiers set out to capture Carter, but Carter single-handedly killed six of the eight and captured the remaining two. Those two captured German soldiers would later provide valuable information to smooth the American advance toward the city of Speyer.

For action above and beyond the call of duty, Edward Carter was awarded the Distinguished Service Cross medal. Carter's commanding officer, however, believed that Carter in fact deserved the Congressional Medal of Honor. But as he saw little chance of a black soldier being given the armed forces' highest honor, he made sure that Carter received the second highest award that could be given a soldier. Later, he was also promoted to Sergeant First Class.

Edward Carter died in 1963 due to lung cancer. He was posthumously awarded the Congressional Medal of Honor in 1997, and the medal was given to Carter's son.

FUN FACT

Carter is also honored within the National Museum of African American History and Culture, as are other African Americans who distinguished themselves during World War II.

CEMETERY

Arlington National Cemetery
End of Memorial Avenue, which extends from the Memorial Bridge
    in Arlington, Virginia
Arlington, Va. 22211
Tel: 877-907-8585
Hours: Daily, 8 AM–5 PM

DIRECTIONS TO GRAVE

From the Welcome Center walk straight until you get to Eisenhower Drive Road and turn left. Go past McClellan Drive to section 59. Turn left for approximately 80 feet. Then take another left for another 60 feet or so, and on the right, you will find Sgt. Carter's gold-leafed gravestone.

AUDIE MURPHY

U.S. Army 2nd Lieutenant
Born: June 25, 1925
Died: May 28, 1971

Audie Leon Murphy was one of the most decorated combat soldiers in World War II, winning every conceivable combat award from the United States, as well as from France and Belgium. He also received the Congressional Medal of Honor for valor at the tender age of 19 in the European theater. After the war, he embarked on a movie career spanning 21 years.

Audie Murphy's Grave

Murphy was born in Kingston, a small rural community in northeast Texas, to Emmett and Rosie Murphy, who were sharecroppers.

As a child he was a loner with mood swings and an explosive temper. He grew up in Celeste, Texas, where he attended elementary school. His father eventually deserted the family and Murphy dropped out of school in the fifth grade. He picked cotton and hunted small game with a rifle to help feed his family. His mom's death in 1941 affected him greatly, and he carried the loss throughout his life.

Murphy always wanted to be a soldier. After Pearl Harbor he tried to enlist, but was turned down because of his age and being underweight. His sister submitted an affidavit that falsified his birth by a year, and he was accepted in the U.S. Army in June 1942. After basic training at Camp Wolters, he was sent to Fort Meade for infantry training, where he honed his marksmanship skills.

He was shipped to Casablanca, Morocco in February 1943. After the surrender of the Axis forces in French Tunisia, his division was put in charge of prisoners. Murphy was promoted to private first class on May 7 and to corporal on July 15. Murphy participated in the mainland Salerno landing in Battaglia, Italy. He then went about distinguishing himself in combat in several situations, killing Germans as well as taking prisoners. In January 1944 he was promoted to staff sergeant. As the result of contracting malaria, he did not participate in the initial beachhead landing in Anzio, Italy, but on January 29 he returned and fought in the first battle of Cisterna. More battle heroics followed with Murphy and his platoon killing the crew of a German tank and then blowing it up.

Murphy participated in the first Allied invasion of southern France. He was with the 15th Infantry regiment during the August 27–28, 1944 offensive at Montelimar that secured the area from the Germans. He was awarded a battlefield commission to second lieutenant on October 14 that elevated him to platoon leader. On October 26, Murphy's platoon was attacked by German troops and he was shot in the hip by a sniper. This kept him out of action until January 1945.

The Colmar Pocket in the Vosges mountains had been held by the Germans since November 1944. His platoon had been transferred there and he rejoined it on January 14, 1945. Murphy was made commander of Company B on January 26. In the town of Holtzwihr they faced a strong German counterattack. Murphy was wounded in both legs, but despite this he single-handedly held off a company of German soldiers for an hour on top of a burning tank destroyer, returning fire to the advancing Germans and killing or wounding 50. He stopped only when he ran out of ammunition and insisted on remaining with his troops till his wounds healed. For Murphy's actions that day, he was awarded the Congressional Medal of Honor. He was promoted to first lieutenant on February 16 and moved from the front lines to regimental headquarters where he was made a liaison officer.

After the war, he was made a captain in the Texas National Guard, where he trained new recruits. He also suffered from PTSD, which caused him insomnia, nightmares, and violent behavior. His first wife, Wanda, whom he had wed in 1949, claimed he once held her at gunpoint. He spoke out about his PTSD to draw attention to it for returning Korean and Vietnam War veterans.

During an acting career that spanned from 1948 to 1969, he made over 40 feature films and one TV series. His most successful film was based on his best-selling book, *To Hell and Back*. The 1955 film was the most successful film made at Universal Studios at the time. He promoted the film on TV with appearances on shows such as the game show *What's My Line* and Ed Sullivan's *Toast of the Town*.

In 1951, after Murphy's divorce from his first wife, he married

Pamela Opal Lee Archer, with whom he had two sons. In the late 1960s he encountered financial difficulties due to his gambling on the horses that he bred and raced at the Del Mar Racetrack. Despite his financial difficulties, he refused to appear in commercials for alcohol or cigarettes, mindful of the influence it would have on young adults.

Murphy was killed on May 28, 1971, when the private plane on which he was a passenger crashed on the side of a mountain near Roanoke, Virginia during inclement conditions and zero visibility. The aircraft was recovered on May 31. Murphy was interred with full military honors at Arlington Cemetery on June 7 of that year.

FUN FACT

Murphy's grave is the second most visited gravesite in Arlington after that of President Kennedy. Unlike other Medal of Honor recipients, his marker does not have gold leaf. Murphy's desire was that his stone be plain, and his family honored his request.

CEMETERY

Arlington National Cemetery
End of Memorial Avenue, which extends from the Memorial Bridge
  in Arlington, Virginia
Tel.: 877-907-8585
Hours: Daily, 8 AM–5 PM

DIRECTIONS TO GRAVE

Murphy's grave is located behind the amphitheater where the Honor Guard marches. It is also in the shadow of the *U.S.S. Maine* monument. By constructing a special flagstone walkway, the cemetery has made it easy to spot, as it is in a small private area in section 46.

VINCENT ROBERT CAPODANNO

Priest/Chaplain
Born: February 13, 1929
Died: September 4, 1967

Vincent Capodanno was born on Staten Island, New York and was the tenth and final child of Italian parents. He received his Bachelor of Sci-

Vincent Robert Capodanno's Grave

ence degree from Fordham University. In 1949 Capodanno entered the Maryknoll Missionary Seminary, and was ordained a priest on June 14, 1958. The following year and until 1965, he was assigned to serve within a parish and school in Taiwan. During that time, he learned to speak Chinese and was able to assimilate within their culture. After a brief return home to the U.S., Father Capodanno volunteered to serve in South Vietnam as a military chaplain. He received his commission as a lieutenant in the Navy Chaplain Corps., and Father Capodanno provided his priestly duties to the 1st Battalion, 7th Marines in South Vietnam. He earned the nickname "the Grunt Padre" because he focused on the spiritual needs of newly enlisted men, or "grunts," as they were called.

Father Capodanno expected no special treatment as a Catholic priest, as he lived, ate and slept under the same conditions as the men to whom he would administer church sacraments. He would always make himself available to console and give advice, spiritual comfort and reassurance whenever he could.

Father Capodanno was severely wounded and killed by enemy fire on September 4, 1967, during his second tour of duty. He was seen running to assist a wounded corpsman, and died as the result of 27 bullet wounds during the North Vietnamese ambush.

On January 7, 1969, he was posthumously awarded our country's highest military honor with the Congressional Medal of Honor. The citation read, "For conspicuous gallantry and intrepidity at the risk of his life above and beyond the call of duty."

In 2002 Capodanno's Cause for Canonization was opened. Then, in 2006, the Catholic Church announced that Father Capodanno had achieved the status of Servant of God. However, it was recently announced in 2022 that it was recommended by a Vatican advisory panel that Capodanno's cause for sainthood be suspended. This was due to an opinion that Capodanno's heroism glorified the military and not the Church. Speaking for ourselves, we hope this opinion is overridden and the cause for Capodanno's sainthood be allowed to continue.

FUN FACT

The *U.S.S. Capodanno*, a frigate, was named in his honor, commissioned in 1973 and blessed by Pope John Paul II in 1981.

THEODORE ROOSEVELT
26th U.S. President
Born: October 27, 1858
Died: January 6, 1919

Theodore Roosevelt is often ranked by historians as among the top five presidents of the United States. He has been described as a force of nature because he lived his life with passion, enthusiasm, and purpose. He was often driven by principle, and this was reflected by his oratorial skills. Even if you disagreed with him, you could not help but be impressed and respect the man.

Theodore Roosevelt was born in New York City to a wealthy family. However, he was often a sickly child who suffered from asthma

Theodore Roosevelt's Grave

so severe that it forced him to sleep sitting up. As he therefore spent much of his time at home, young Theodore developed into a voracious reader and evolved an interest in zoology. This interest he carried into adulthood.

Theodore Roosevelt was greatly influenced by his father, Theodore Roosevelt, Sr. The future president described his father as a kind man who nevertheless would not tolerate from any of his children failing to do their best at anything they attempted. He also instilled in them the virtue of unselfishness.

Young Theodore soon discovered that physical exercise would relieve many of his asthma symptoms. He took boxing lessons after an incident when two bullies roughed him up due to his frail appearance.

Theodore graduated magna cum laude from Harvard in 1880. In the same year, he married his first wife, Alice Lee. In 1882 Roosevelt had his first book published, entitled *The Naval War of 1812*. Also in 1882, he was elected to the first of three terms in the New York State Assembly, where he distinguished himself as a reformer determined to rout corruption. Roosevelt and his wife had one daughter, also named Alice, on February 12, 1884. However, two days later, Roosevelt's wife died from complications arising from her pregnancy. Further, earlier that very day, Theodore suffered the loss of his mother from typhoid fever. As to be expected, these personal double tragedies affected Theodore deeply. For the rest of his life, he rarely spoke or wrote of his first wife.

In the prior year of 1883 Theodore had purchased ranch land in the Bad Lands of the Dakotas. After his wife's and mother's deaths he decided to try his hand at raising cattle. Though he enjoyed the life of being a genuine cowboy, Theodore decided to sell his ranch at a loss following a severe winter that had wiped out his herd.

In 1886, at the tender age of 28, Theodore Roosevelt ran for mayor of New York City. He was defeated as the Republican nominee running third in a three-way race. Following his loss, in December 1886 he married his second wife, Edith Carow, with whom he had five children.

President Benjamin Harrison appointed Roosevelt to head the U.S. Civil Service Commission in 1889. Roosevelt fought the corrupt spoils system, demanding that civil service laws be observed. Despite Theodore being a Harrison supporter, when Grover Cleveland won the presidential election of 1892, Cleveland reappointed Roosevelt to the same position.

In 1894, Roosevelt was offered and accepted the position of police commissioner for New York City. He was known for walking at night to check up on police officers and see if they were doing their job patrolling their beats. Woe to any officer who was found drinking on the job, as they would receive a tongue-lashing from Roosevelt that they would not forget.

In March of 1897, Roosevelt accepted President McKinley's offer to be the assistant secretary of the Navy. The following year, upon the sinking of the *U.S.S. Maine*, the Spanish American War began when it was believed that Spain bore the responsibility for the sinking of the *U.S.S. Maine*. Theodore volunteered to lead a troop of men, now known as the Rough Riders, in Cuba. In battle Roosevelt distinguished himself with his courage and daring. Rough Riders took San Juan Hill in Cuba, and shortly thereafter Spain gave up their hostilities in Cuba. In 2001 Theodore was posthumously awarded the Congressional Medal of Honor for this effort.

In 1898, Roosevelt sought the Republican nomination for governor of New York State and won the caucuses easily. In November of that year, Theodore Roosevelt won the governorship in a close race. Just two years later, he found himself on the national ticket as William McKinley's running mate for the vice presidency. McKinley was easily reelected. However, just a little more than six months later, President McKinley was assassinated, and Theodore Roosevelt at age 42 became the nation's 26th president.

Roosevelt became known as a trust-buster. He did not hate big business, but when he saw what big business was capable of—such as colluding to charge higher prices and restraining fair trade—Roosevelt would apply, when applicable, the Sherman Anti-Trust Act.

When Japan and Russia declared war against each other in 1904, they requested that Roosevelt mediate an international conference among themselves. For his efforts, Theodore Roosevelt became the first U.S. president to win the Nobel Peace Prize.

Roosevelt was determined to lead the build-up of the U.S. Navy,

Letter signed by Roosevelt, from the private collection of Robert Gardino

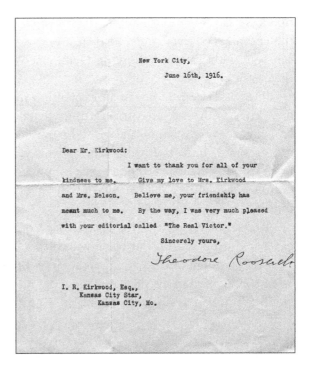

and by the end of his presidency the Navy had increased the number of battleships to the point that only Great Britain had more. Theodore Roosevelt is famous for the adage "Speak softly but carry a big stick," and as an example of practicing what he preached, Roosevelt proceeded to send many of the new battleships on a world tour.

The biggest accomplishment of the Roosevelt administration was the construction of the Panama Canal, which was completed in 1914 under the Taft presidency. When the canal became operational, U.S. warships, as well as commercial vessels, could then move at a much more rapid rate from the Pacific to the Atlantic Ocean and vice versa.

In 1912, Theodore Roosevelt was disappointed in his hand-chosen successor, William Howard Taft, as the Taft administration pursued more conservative policies that were viewed by Theodore as a slight to Roosevelt's more progressive legacy. Roosevelt unsuccessfully challenged Taft for the Republican party nomination, so Roosevelt then ran as a third-party candidate of the popularly labeled Bull Moose party. (Roosevelt meanwhile survived an attempted assassination less than one month before the November election.) However, in the end all Roosevelt succeeded in doing was splitting the Republican vote with Taft and allowing Democrat Woodrow Wilson to win the election.

In 1913, Theodore began an expedition into the Brazilian wilderness. His health took a disastrous turn when he developed jungle fever, which is a severe form of malaria. This was exacerbated from complications due to infection from the previous year's attempted assassination. Roosevelt survived, but never fully recovered. The tragic death of his son Quentin in a plane crash in France only worsened his health difficulties.

Roosevelt was critical of Woodrow Wilson's handling of America's role during World War I. He voiced that he was amenable to running for president once again in 1920. However, it was not meant to be, as Theodore died in his sleep on January 5, 1919, as the result of a blood clot that traveled to his lungs.

FUN FACTS

Theodore Roosevelt was the most successful third-party candidate in the 20th century. In the election of 1912, he came in second place in the popular vote and won the most electoral votes (88) versus any other major third-party presidential candidate.

Our 26th president has 26 steps that must be climbed to get to his gravesite.

CEMETERY

Young's Memorial Cemetery
134 Cove Neck Road
Oyster Bay, NY 11771
Tel.: 516-922-4788
Hours: Daily, 9 AM to dusk

DIRECTIONS TO GRAVE

Upon entering the cemetery, take the paved path up, going straight past three benches. You will see a stairway on the left. Take that stairway to the gravesite.

RICHARD BYRD
Aviator/Polar Explorer
Born: October 25, 1888
Died: March 11, 1957

Richard E. Byrd's heroic exploits both in flight and in exploration enthralled the American public in the 1920s and '30s. Aircraft flights that he commanded as a navigator or expedition leader crossed the Atlantic Ocean, a segment of the Arctic Ocean and a segment of the Antarctic Plateau. The expeditions that he led were the first to reach the North and South Poles by air.

Byrd was born in Winchester, Virginia, the son of Esther and Richard Evelyn Byrd, Sr. He attended the Virginia Military Institute before transferring to the University of Virginia. He ultimately was appointed to the U.S. Naval Academy, where he graduated on June 8, 1912, commissioned as an ensign. He got an assignment on the gunboat *U.S.S. Dolphin*, which served as the yacht for then Secretary of the Navy Josephus Daniels. There he met many influential men such as Franklin Roosevelt. He rose to the rank of lieutenant junior grade on June 8, 1915. He was later assigned to the presidential yacht the *U.S.S. Mayflower*, and there he was retired, due to an ankle injury suffered on board on March 15, 1916.

On January 20, 1915, he had married Marie Donaldson Ames. The union produced four children. They settled into a large brownstone house on Brimmer Street in the Beacon Hill section of Boston, where they lived for the rest of their lives.

Shortly after the start of World War I, Byrd oversaw the mobilization of the Rhode Island Naval Militia. He qualified as an aviator and commanded naval air forces in Nova Scotia. For his services during the war, he received a letter of commendation from Secretary of the Navy Josephus Daniels.

After the war, in 1919, he volunteered to be a crew member in the U.S. Navy's historic first aerial trans-Atlantic crossing. He planned the flight path of the successful flight of Lieutenant Commander Albert Read on May 18, 1919.

Byrd also volunteered to fly across the Atlantic solo, but this was quashed by Secretary of the Navy Theodore Roosevelt, Jr., who deemed it too dangerous.

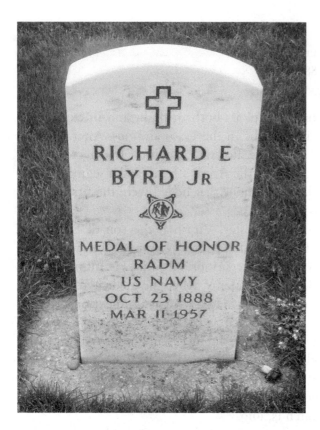

Richard Byrd's Grave

During his command of the arctic exploration in Northern Greenland in 1925, he met Navy Chief Aviation Pilot Floyd Bennett and Norwegian pilot Bennett Balchen. Bennett served as his pilot to the North Pole the next year and Balchen was the primary pilot on Byrd's flight to the South Pole in 1929.

On May 9, 1926, Byrd and Bennett attempted a flight over the North Pole in a Fokker tri-motor monoplane, and successfully completed the trip in just under 16 hours. On his return, Byrd became a national hero and was promoted to commander in addition to receiving the Medal of Honor from President Calvin Coolidge.

In 1927, Byrd was aced out of crossing the Atlantic nonstop by Charles Lindbergh due to the fact that his Fokker airplane crashed in a practice takeoff. The delay to repair it allowed Charles Lindbergh to be the first to cross the Atlantic. Byrd's flight finally reached France on July 1, 1927. On board he carried the U.S. mail to demonstrate the practicality in addition to the drama of aircraft.

In 1928, Byrd embarked on the first Antarctic expedition, involving two ships and three airplanes. Photographs and geological surveys were taken during the summer. On November 28, 1929, the first flight over the South Pole was successfully undertaken with Balchen as the pilot. This made Byrd even more famous not only in the U.S. but internationally.

On his second expedition to the Antarctic in 1934, Byrd spent five winter months alone operating a meteorological station. He nearly lost his life due to carbon monoxide fumes from a poorly ventilated stove. He was rescued by his men, who were finally able to reach him under the harsh conditions. He wrote a best-selling book, *Alone*, about his experience.

His third expedition to the Antarctic, financed by the U.S. government from 1939 to 1940, focused on extensive studies of geology, biology, meteorology and exploration. He was recalled to active Navy duty as a senior officer and became a confidential advisor to Admiral Ernest King for the duration of the Second World War. He was present at the Japanese surrender at Tokyo Bay on September 2, 1945.

In 1946, Secretary of Defense James Forrestal appointed Byrd officer in charge of the Antarctic development project, code-named Operation High Jump. This was the largest Antarctic expedition to date with over 4,000 personnel involved. It made aerial explorations of an area half the size of the United States, recording 10 new mountain ranges. In 1948, the U.S. Navy produced a documentary of the expedition entitled *The Secret Land*.

In 1955, Byrd was appointed officer in charge of the U.S. Navy Operation Deep Freeze I. It established permanent Antarctic bases in McMurdo Sound, the Bay of Whales and the South Pole. This marked the beginning of a permanent U.S. military presence in Antarctica. It was his last trip there, and he remained only one week, returning to the United States on February 3, 1956.

Byrd's health suffered a serious decline upon his return with heart problems. He died in his sleep as a result of a heart ailment at his home in Boston on March 11, 1957. He was interred with full military honors at Arlington National Cemetery.

FUN FACT

Byrd actively cultivated the friendships of powerful and important people such as Franklin Roosevelt, Henry Ford, Edsel Ford, John D. Rockefeller, Jr. and Vincent Astor. They provided the necessary political and financial support for his expeditions. In gratitude, he named geographic features such as mountain ranges after them.

CEMETERY

Arlington National Cemetery
End of Memorial Avenue, which extends from the Memorial Bridge
    in Arlington, Virginia
Tel.: 877-907-8585
Hours: Daily, 8 AM–5 PM

DIRECTIONS TO GRAVE

Turn right on sidewalk walkway after the Welcome Center. Bear right to stay on. Bear left on crosswalk. Turn left on Roosevelt Drive. Turn right on paved sidewalk. Turn right on Grant Road. Bear left on paved walkway. Make a sharp left on grave row intersection. Turn right to stay on grave row intersection. Byrd will be on the left.

ALVIN YORK

Soldier/Sergeant
Born: December 13, 1887
Died: September 2, 1964

Despite initially claiming to be a conscientious objector due to his religious beliefs, Alvin York became one of the most decorated U.S. soldiers in World War I. He received the Medal of Honor for leading an attack on a German machine gun nest, killing at least 25 Germans and taking 132 prisoners. France, Italy and Montenegro bestowed decorations on him.

Alvin Cullum York was born in a two-room log cabin in Fentress County, Tennessee to William and Mary York. The family was impoverished, and William worked as a blacksmith. The York sons attended some school, but soon left to work on their farm to help feed

Alvin York's Grave

the family. When William died in 1911, Alvin pitched in to help his mother raise his siblings since he was the oldest. He worked in railroad construction and as a logger. He was known to be a skilled worker devoted to his family. He was also an alcoholic who got into many saloon fights, one of which saw his best friend killed.

Despite his alcoholism, he attended church regularly. A revival meeting at the end of 1914 led to a conversion experience on January 1, 1915. York's church, the Church of Christ in Christian Union, a Presbyterian denomination, had a strict doctrine of pacifism. Conscientious objectors then could still be drafted and given assignments that didn't conflict with their principles. In November 1917, his application was considered, and he was drafted. York served in G Company, 328th Infantry, 82nd Division. Deeply troubled by his religious beliefs, he spoke to both his company commander and battalion commander. Biblical passages about war persuaded York to reconsider the morality of his participation in the war. After a 10-day leave, York returned, convinced that God wanted him to fight and would keep him safe.

In an October 8, 1918 attack during the Meuse-Argonne offensive, Private (acting corporal) York's actions were outstanding. York was with a group of 17 soldiers assigned to infiltrate German lines and silence a machine gun position. After the patrol had captured a large number of enemy soldiers, German small-arms fire killed six Americans and wounded three. Several Americans returned fire while the others guarded the prisoners. York and other soldiers attacked the machine gun position, killing several German soldiers. The German officer responsible for the machine gun position emptied his pistol shooting at York, but failed to hit him. The officer then offered to surrender, and York accepted, taking 132 prisoners. York was promoted to sergeant and received the Distinguished Service Cross, which was later upgraded to the Medal of Honor. He eventually received 50 decorations. As one would suspect, his feat made him a national hero who was admired throughout the world.

Upon his return stateside and his discharge on June 7, 1919, York married Gracie Loretta Williams. They would have ten children.

A group of Tennessee businessmen purchased a farm for York that was not fully equipped, and York had to stock it himself by borrowing money. Following the post-war economic depression, the Rotary could not make the payments, which thus devolved upon York. Only a plea for public subscription, and a story on his plight by the newspaper *New York World*, saved the farm by Christmas 1921.

In the 1930s and '40s York worked as project superintendent for the Civilian Conservation Corps. With the onset of World War II, York abhorred isolationism and wanted to enlist again to fight. His age and maladies worked against him, and he was turned down. He was commissioned a major in the Signal Corps and toured Army camps, fundraising with bond drives.

For many years, York had resisted making a film about his exploits, but in 1940, wanting to finance an interdenominational Bible school, he acquiesced. The 1941 movie *Sergeant York*, starring Gary Cooper, earned Cooper the Academy Award as best actor. York earned $150,000 after two years as well as later royalties. This enabled him to build his Bible school.

Starting in the 1920s, York had many ailments, from gallbladder surgery to pneumonia. In 1948, he suffered a stroke. More strokes followed, and he was confined to a bed, which lasted a lengthy period of time until his death. York died at the Veterans Hospital in Nashville, Tennessee on September 2, 1964, of a cerebral hemorrhage at the age of 76. With General Matthew Ridgeway representing President Lyndon Johnson, he was buried at the Wolf River Cemetery in Pall Mall, Tennessee.

FUN FACTS

After bringing his prisoners back to the American lines in the war and returning to his unit, he reported to his brigade commander, Brigadier General Julian Robert Lindsey. Lindsey remarked, "Well York, I hear you have captured the entire German army." York replied, "No sir. I got only 132."

Gracie and Alvin named quite a few of their children after historical figures such as Thomas Jefferson, Woodrow Wilson, Andrew Jackson, and Betsy Ross.

CEMETERY

Wolf River Cemetery
3500 Wolf River Loop
Pall Mall, Tennessee 38577
No phone number is available. The cemetery is open 24 hours.
Email: wolfrivercemeteryassn@gmail.com

DIRECTIONS TO GRAVE

Off the Wolf River Loop Road, enter the cemetery at the second entrance, and you'll see York's grave under the American flag.

CHAPTER TWO
U.S. Presidents

CHESTER ALAN ARTHUR
21st U.S. President
Born: October 5, 1829
Died: November 18, 1886

Modern historians generally describe Arthur's presidency as mediocre—one of the least memorable. That glosses over the fact that when he left office, he received praise from his contemporaries for his solid performance. That was the result of his courageous actions and initiatives when he had taken over the reins of government.

Born in Fairfield, Vermont to parents of Scotch and Irish descent, Arthur grew up in Schenectady, New York and attended Union College, where he studied the traditional classical curriculum and was elected to Phi Beta Kappa. Upon graduation and a move to New York City, Arthur studied law at the office of Erastus Culver. Arthur joined this firm after being admitted to the New York bar in 1854. In 1856, he married Ellen Herndon, the daughter of a Virginia naval officer. They had three children together.

At the onset of the Civil War, Arthur was appointed to the post of brigadier general. In July 1862, Arthur became quartermaster general.

Toward the end of the war, Arthur became involved in New York Republican politics, which was then dominated by party leader Thurlow Weed. Arthur aligned himself with Roscoe Conkling, a rising upstate congressman. In 1870, President Grant gave Conkling control over New York patronage in gratitude for his fund-raising efforts in getting Grant elected in 1868. Conkling then appointed his friend Arthur to the prestigious and lucrative position of the collector of the Port of New York. As such, he controlled over 1,000 jobs and oversaw the awarding of many city contracts.

Arthur's tenure came to an end with the election of President Hayes in 1877 and Hayes' pledge to reform the patronage system. Hayes terminated him in 1878. Arthur then became involved in New York City politics and eventually became chairman of the Republican Executive State Committee.

In January 1880, his wife, Ellen Arthur, fell seriously ill with pneumonia while Arthur was in Albany attending to organizing the political agenda for the coming year. He barely made it back to New York

Chester Alan Arthur's Grave

City to see her alive for the final time. When Ellen died, Chester was devastated, and he never remarried. He kept a photo of Ellen on his desk at the White House, where he placed flowers daily during his tenure there.

At the 1880 Republican convention, Arthur was offered the vice-presidential nomination as the running mate of James Garfield. Garfield knew he needed the support of New York Republicans to win, which he narrowly did. Despite entreaties by Conkling not to accept the vice presidency, Arthur did, saying it was a great honor.

However, Garfield and Arthur did not see eye to eye. He did not appoint Conkling/Arthur supporters, known as Stalwarts, to major cabinet positions. He finally relented on postmaster general to appease the Stalwarts, appointing Thomas James.

Everything changed on July 2, 1881, when a disgruntled office-seeker, Charles Guiteau, shot Garfield in the back as he strode arm in arm with James Blaine, Garfield's secretary of state, at the Washing-

Signature of Chester Alan Arthur, from the private collection of Robert Gardino

ton, DC train station. Garfield lingered for 80 days during that summer until September 19, when he expired from his wounds.

Arthur was sworn in as president at 2:15 AM on September 20 by Judge John R. Brady of the New York Supreme Court. Arthur's dignified behavior during Garfield's ordeal softened the public's attitude toward him. Roscoe Conkling and the Stalwarts rejoiced, thinking they had someone whom they could manipulate to do their bidding. They were in for a surprise, however, as Arthur courageously turned the tables on them and became a reformer.

He advocated for and enforced the Pendleton Civil Service Act. By 1884, half of all postal officials and three-quarters of customs jobs were awarded by merit. He presided over the rebirth of the U.S. Navy, which had declined precipitously after the Civil War. He created an advisory board to build a Navy to protect America—thousands of miles away and not just in our coastal waters. Arthur's efforts put into motion the construction of eleven steel warships, which were completed by 1889.

Arthur was diagnosed with Bright's disease right after he became President. This disease, which is ultimately terminal, severely affected his kidneys and debilitated Arthur during his presidency. His sister Mary Arthur McElroy, who was the acting First Lady, tenderly took care of her brother's health issues at the White House as he courageously sought to perform his duties.

Due to Arthur's illness, he only half-heartedly sought the GOP nomination for President in 1884 and returned to New York City to practice law at the conclusion of his term. He finally succumbed to his disease on November 18, 1886, aged only 57 years. He was laid to rest beside his beloved wife, Ellen, in Albany's Rural Cemetery, located in Menands, New York.

FUN FACT

Arthur began corresponding with Julia Sand when he was Vice President and Garfield lay dying. A young woman from a well-to-do family in New York, he had not known who she was prior to receiving her letters. She encouraged him to seize the office of the Presidency with vigor upon Garfield's death. An invalid, she invited Arthur to visit her at her home in New York City. He surprised her and did so on August 20, 1882, staying an hour and raising her spirits immeasurably. He kept the 23 letters that she sent him in a special large envelope that was discovered by his son after his death. He referred to her as his conscience.

CEMETERY

Albany Rural Cemetery
3 Cemetery Avenue
Albany, New York 12204
Tel.: 518-463-7017
Hours: 8:30 AM–4:30 PM

DIRECTIONS TO GRAVE

Enter cemetery at the Cemetery Avenue entrance. The office will be on your left. Continue left and take the road, which is Ridge Road, straight up. When the road shifts to the left, follow it to Arthur's grave. His grave has an American flag flying, which is easy to spot.

HERBERT HOOVER
31st U.S. President
Born: August 10, 1874
Died: October 20, 1964

Today when the Great Depression of the 1930s is brought up, most people will associate Herbert Hoover's name with it. Also, most will blame his administration for it. To be fair, an examination of the facts

Herbert Hoover's Grave

will bear out that Hoover did not cause the Great Depression. He was elected president of the United States in November of 1928 and sworn in as president on March 4, 1929. The severe economic turmoil that would last over a decade began with the stock market crash in October 1929, just months after Hoover was sworn into office. Therefore, we consider it a gross injustice to blame the Great Depression on anything Hoover did in his first few months in the oval office. Certainly, however, Hoover can be blamed for pursuing economic policies that in hindsight were wrong and may have made the Depression worse. We discuss this in more detail in his biography that follows. But our failure to include Herbert Hoover in this book would be a great disservice to a man who was primarily responsible for feeding millions of starving people during World War I. For his humanitarian efforts, Hoover was nominated for the Nobel Peace Prize a total of five times.

Herbert Hoover was born in Cedar County, Iowa in 1874 to a Quaker family. Both of his parents died by his tenth birthday. He was taken in by an uncle and moved to Oregon. After graduating from Stanford University, he became a mining engineer. He became so successful that Hoover started his own engineering consulting firm. By the time he was in his 30s, he was a multimillionaire. With the outbreak of World War I, Hoover founded the international Commission for the Relief of Belgium. The purpose of this organization was to protect and provide travel to over 100,000 Americans who were stranded when Germany invaded Belgium. After creating the Federal Food Administration (later renamed the American Relief Administration) President Woodrow Wilson appointed Hoover to head it. Hoover is credited with providing food to millions of starving central and eastern Europeans when he headed the American Relief Administration after the war. He took criticism for providing food even to the Russians after the Bolsheviks had taken over that government. Hoover retorted, "Twenty million people are starving. Whatever their politics, they shall be fed!"

In 1921, President Warren Harding chose Hoover as his secretary of commerce. In this cabinet position he served under both Harding and his successor, Calvin Coolidge. Hoover was a highly visible member of the cabinet and was widely admired for his efficiency and competence. He was the most obvious choice for the Republican party in the race for the White House in 1928. Hoover won an overwhelming victory over the Democratic nominee, Governor Alfred Smith of New York. Hoover won approximately 58 percent of the popular vote and carried 40 of the then 48 states. However, the period in the U.S. known as the "roaring 20s" ended abruptly when the stock market crashed in October 1929.

Most economists attribute the market crash to the Federal Reserve when the Fed began to tighten credit to curb what they saw as excessive stock market speculation. The Fed began to raise its discount rate, which is the rate the Fed charges member banks to borrow directly from the Fed. By contracting the money supply too much, this in turn led smaller banks to fail at an alarming rate. Depositors lost their money and businesses could no longer easily borrow. In 1930, Hoover also made the mistake of signing the international Smoot-Hawley Tariff Act. By forcing countries to pay higher fees to be able to sell their goods in the U.S., many were led to retaliate in kind against the United States. Spain, for example, increased tariffs by 150 percent on American cars. This in essence shut American autos completely out of the Spanish market. Finally, another major error committed by Hoover was increasing the personal tax rate via the Revenue Act of 1932. Hoover did this because of his belief that federal budget deficits would be more harmful in the long run than increased taxes.

In 1932, the unemployment rate in America rose to an unprecedented 23.6 percent. Bread lines became a common sight. Homelessness grew, and shack towns and encampments became known as Hoovervilles. As a result, Herbert Hoover, who four years before had been elected in a landslide, found himself defeated in a landslide by Democratic nominee Franklin Roosevelt. FDR received approximately 57 percent of the popular vote, winning 43 out of 48 states. Franklin Roosevelt would remain in office for slightly over 12 years. Hoover spent much of that time defending his administration record as well as being a critic of FDR's New Deal economic policies, and also his foreign policies.

Following our victory in World War II, in 1946 President Harry Truman chose Hoover to tour both occupied Germany and Italy, and to see to their food needs. Hoover made sure that millions of children in those countries were fed. The following year, Truman chose Hoover to lead the Commission on Organization of the Executive Branch of the Government. From this "Hoover Commission" recommendations were made on eliminating waste, fraud and inefficiencies, consolidating agencies and strengthening White House control over policies.

Herbert Hoover died at age 90 in October 1964 from internal bleeding as the result of intestinal surgery for the removal of a tumor.

FUN FACT

In an ultimate presidential irony, Franklin Roosevelt, who would go on to defeat Hoover for the presidency in 1932, said in 1920 of Herbert Hoover, "I wish we could make him President of the United States. There could not be a better one."

BURIAL

Herbert Hoover is buried at the Herbert Hoover Presidential Library and Museum, which is open daily from 9 AM to 5 PM. There is an entry fee of $10 per adult. The site is located at:

210 Parkside Drive
West Branch, Iowa 52358

CHAPTER THREE
U.S. Government Officials

WILLIAM SEWARD
Secretary of State/Governor/Senator/Statesman
Born: May 16, 1801
Died: October 10, 1872

When the name of William Seward is mentioned, most people will probably remember him as the man responsible for the purchase of Alaska, a purchase mockingly tagged at the time as "Seward's Folly." Seward was a top member of Abraham Lincoln's cabinet during the Civil War. Need we say more on why we think Seward is rightly included in this volume?

William H. Seward was born in Orange County, New York, and was an exceptional student. After completing college with highest honors, Seward studied law, and at the age of 21 passed the bar exam. After years of practicing law, Seward was drawn to politics. In 1839 he was elected governor of New York, and was elected to the U.S. Senate in 1849. Seward took strong stands against slavery and was reelected to the Senate in 1855. Initially a member of the Whig party, Seward joined the then newly formed Republican party. Going into the Republican convention of 1860, held in Chicago, he was the favorite to win the Republican nomination for president. Seward in fact led on the first ballot, but fell short of the required number of votes to be the nominee. He eventually lost to Abraham Lincoln on the third ballot.

Despite Seward's disappointment at losing to Lincoln, his support for Lincoln in the November 1860 election was pivotal in Lincoln carrying New York State and thus winning the election. Lincoln then chose Seward to be his secretary of state. In this cabinet position, Seward demonstrated adept diplomatic skills, as he was able to dissuade both Great Britain and France from directly supporting and recognizing the Confederacy as a separate nation.

Shortly after Lincoln's second inauguration in March 1865, Shakespearian theater actor John Wilkes Booth plotted not only to assassinate Lincoln, but, with fellow conspirators, to kill Vice President Andrew Johnson and Secretary of State Seward as well. Booth chose George Atzerodt to kill Johnson and Lewis Powell to kill Seward. Atzerodt turned out to be a simpleton who chickened out of attempting to kill Johnson. Lewis Powell, however, was a tall and physically

William Seward's Grave

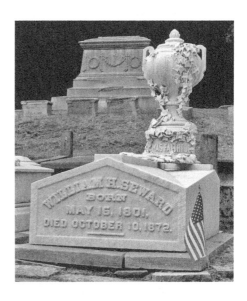

imposing man, and on the night Booth shot Lincoln in Ford's Theater, Powell arrived at the Seward residence to do his assigned deed. A few days earlier, Seward had been involved in a major accident involving his horse carriage. Seward required a collar brace and chest brace. Powell gained entry to Seward's home on the ruse that he was delivering medicine to the ailing Seward. Once inside the residence, Powell overpowered Seward's son Frederick and daughter Fanny. He then leaped on to the bed where a helpless Seward was lying and stabbed him several times. The collar and chest brace ultimately saved Seward's life, but not before Powell's knife did major damage to his face. Seward would remain disfigured for the rest of his life. Lewis Powell was caught and, along with the other Booth co-conspirators, was hanged on July 7, 1865.

Seward had meanwhile long been a proponent of expanding American territories. In 1867, Seward purchased the territory of Alaska from Russia for $7,200,000. History has since shown that Alaska is rich in natural resources, including oil and natural gas. Of course, additionally, no one knew at that time the strategic importance of Alaska being close to Russia. The acquisition was far from a folly!

William Seward died in 1872 at age 71 from complications of the attempt made on his life seven years earlier.

FUN FACT

Besides acquiring the Alaskan territory, Seward expressed a desire to purchase also Greenland and Iceland. The idea of purchasing Greenland has additionally been proposed by President Harry Truman and, more recently, President Donald Trump.

CEMETERY

Fort Hill Cemetery
19 Fort Hill Cemetery
Auburn, NY 13021
Tel.: 315-253-8132
Hours: The grounds are typically open from sunrise to sunset.

DIRECTIONS TO GRAVE

Enter the cemetery from the West Lake Avenue entrance. Follow the road past Bradley Chapel Memorial Office and continue on the road past the garage. Take it to Fort Road and then take a right. Follow curved road past Buzz Circle. Follow this road as it curves, always bearing left until you get to a loop called Glen Cove. In front of the point at the Glen Cove section is the Glen Haven section with a sign that says Glen Haven 14. Seward's grave is up the small steps on the left. His and his wife's grave are topped by urns.

FIORELLO H. LA GUARDIA
New York City Mayor
Born: December 11, 1882
Died: September 20, 1947

Fiorello H. La Guardia is still the gold standard by which all U.S. mayors are judged. He was ranked number one by 69 scholars in 1993 as the best mayor in American history. He achieved that stature by resurrecting the floundering city left to him by his predecessor and turning it around during one of the most difficult periods that America has endured, namely the Great Depression. La Guardia consolidated New York City's transit system, making it more efficient. He also expanded construction of public housing, playgrounds, parks, and airports while reorganizing the New York City Police Department, and

Fiorello H. La Guardia's Grave

further, he curbed the power of the Irish-controlled Tammany Hall political machine.

Fiorello Raffaele Enrico La Guardia was born in Greenwich Village, New York to Achille and Irene La Guardia. His father, a former Catholic, was an atheist and his mother, a non-practicing Jew. His father had enlisted in the Army as a warrant officer and a musician in 1885, and the family moved around a bit, living in the Dakota Territory and Prescott, Arizona, where Fiorello was enrolled in the Episcopal church, a religion he practiced all his life.

After returning to America from Europe, Fiorello worked at a series of odd jobs as a translator and clerk. He graduated from Dwight School on the upper west side of New York City. He earned his law degree from NYU and was admitted to the bar in 1910. He joined a Republican club while at NYU. In 1919, he ran in a special election for president of the New York City Board of Aldermen and won. He sought the Republican mayoral nomination, but was defeated in the primary.

Fiorello married Thea Almerigotti, an immigrant from Trieste, Italy, in 1919 in a Catholic ceremony at St. Patrick's Cathedral. Tragedy soon marred their happiness, however, with their child Fiorella Thea (born in June 1920) dying the following May, followed by Thea herself that November. Fiorello was shattered and immersed himself in his work to cope with "the greatest tragedy in his life." He eventually married his secretary, Marie Fisher, in a Lutheran ceremony in 1929.

Briefly a member of the socialist party in the early 1920s, La Guardia returned to the Republican party in 1926, winning a House seat by a narrow margin (55 votes); he was the only Republican elected from New York City. He had wanted to run for mayor in 1925, but declined because he thought he would lose to Mayor Jimmy Walker. He did run for mayor in 1929 but was defeated in a landslide when he ran against Walker.

With Mayor Jimmy Walker forced out of office by the Samuel Seabury Investigation (a reform movement), Fiorello ran in 1933 on a platform ticket of the Fusion Party, which was made up of Republicans, reformed-minded Democrats and Independents. It promised honest government with greater effectiveness and inclusiveness. The support of reformer Samuel Seabury was a big boost to his candidacy, and he won the first of three mayoral terms, becoming New York City's 99th mayor.

During those three terms, which encompassed the Great Depression and World War II, La Guardia accomplished much. He restored fiscal health to the city, extended the federally-funded work relief program to the unemployed, ended city corruption and racketeering in key sectors, instituted a merit-based civil service and modernized the infrastructure with respect to city parks and transportation. He also waged a war against crime with the aid of special prosecutor Thomas Dewey, zeroing in on the mob bosses Lucky Luciano and Frank Costello. He famously rounded up thousands of slot machines, known as "one-armed bandits," and, swinging a sledgehammer, demolished them before dumping them in the river.

By 1944, unable to secure funds from FDR for his programs, his popularity waned and he slipped in straw polls. This forced him to decline to run for a fourth term. President Truman made him the head of the United Nations Relief and Rehabilitation Administration

(UNRRA) in early 1946, but the UNRRA shut down by the end of 1946.

Diagnosed with inoperable pancreatic cancer in early 1947, he succumbed to the disease on September 20 at his home in the Riverdale section of the Bronx at the age of 64. After lying in state with a funeral at the Cathedral of St. John the Divine, he was buried in Woodlawn Cemetery in the Bronx.

FUN FACTS

La Guardia enlisted and fought in World War I as an aviator. He trained pilots in Foggia, Italy. Rising to the rank of major, he was presented the Flying Cross by King Victor Emmanuel III of Italy in 1917.

Because of his small stature, standing only 5'2", La Guardia was given the nickname of "the Little Flower."

During the newspaper strike of July 1945, La Guardia famously read the comics on the air over WNYC Radio.

CEMETERY

Woodlawn Cemetery
4199 Webster Avenue
Bronx, New York 10470
Tel.: 718-920-0500
Hours: Daily, 8:30 AM–4:30 PM

DIRECTIONS TO GRAVE

Enter at Webster Avenue entrance and as you enter follow Central Avenue. Turn left when you hit Myosotis Avenue. Make another left on Alpine Avenue and Fiorello's grave will be right off the road about 25 feet on the right.

ALEXANDER HAMILTON
Founding Father/Statesman
Born: January 11, 1757
Died: July 12, 1804

Alexander Hamilton belongs to the group of Founding Fathers who established our republic. As the first secretary of the treasury, he set the newborn country on a path of fiscal soundness. His untimely death in a duel at an early age ended a career that was packed with contributions to the new nation.

Hamilton was born out of wedlock to Rachel Foucette, a married woman who was half British and half Huguenot, and James Hamilton, a Scotsman, on January 11, 1757 in Charlestown, the capital of the Caribbean island of Nevis. Orphaned early, he was given a home by Thomas Stevens, a merchant also from Nevis. As a teenager, he became a clerk at a local import/export firm and was left in charge for five months while the owner was at sea. A detailed letter to his father about a devastating hurricane that hit the island impressed community leaders, who collected a fund to send Hamilton to the North American colonies to be educated.

Landing in Boston in 1772, he proceeded to New York City where he enrolled in King's College, now Columbia University. Unfortunately, he was forced to discontinue his studies during the British occupation of the city. In 1775, Hamilton, along with other students, joined a volunteer militia. Through his influential connections with New York patriots, he raised the New York Provincial Company of Artillery of 60 men and was elected captain. They saw action against the British at the Battles of Trenton, Harlem Heights, White Plains, and Princeton. While stationed in Morristown, he met and fell in love with Elizabeth Schuyler, a daughter of General Philip Schuyler. They were married at the Schuyler mansion in Albany on December 14, 1780, and had eight children together.

He was extended an invitation to serve General Washington as chief of staff with the rank of lieutenant colonel. He served ably for four years, but he yearned for a field command. Washington finally relented, and Hamilton saw combat at the climactic siege of Yorktown.

After the war, he served as a New York delegate to the Congress of Confederation in 1783. He resigned the following year and founded the Bank of New York. In 1788 he was awarded a MA from Columbia after playing an integral part in reopening the college.

After serving as an assemblyman from New York County, he was chosen as a delegate to the Constitutional Convention in Philadelphia. Despite having problems with the original draft, he signed it anyway, seeing it as an improvement over the Articles of Confederation. He

Alexander Hamilton's Grave

was the only New York signer, and helped play a role in its national ratification. This was done with the aid of "the Federalist Papers," a series of essays designed to defend the proposed Constitution. Hamilton supervised the entire project and recruited John Jay and James Madison to help. He made the largest contribution to the effort, writing 51 of the 85 papers and signing them "Publius." The Constitution was ratified on May 29, 1790, after Rhode Island finally affirmed it.

In the newly formed cabinet of President Washington, he became the first secretary of the treasury. Hamilton envisioned a central government with a strong national defense and industrial economy. He argued that the implied powers of the Constitution provided the legal authority to fund the debt, as well as assuming the states' debt, and create the First Bank of the United States. The Coinage Act of 1792 established the U.S. Mint. His views formed the basis of the Federalist Party, which was in contrast to the Democratic-Republican Party, headed by Thomas Jefferson, which favored an agrarian economy.

After resigning as treasury secretary, he resumed his legal and business activities and was a leader in the movement for the abolition of the international slave trade.

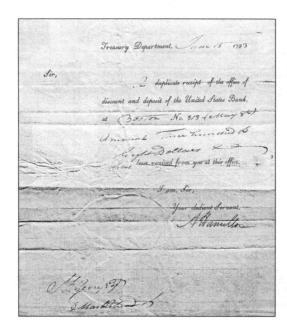

Hamiton's signature on a check, from the private collection of Vincent Gardino

In the 1804 Presidential election, he opposed Adams' reelection. Jefferson and Aaron Burr tied for the presidency, and feeling Jefferson was the lesser of two evils, Hamilton threw his support behind him, and he was thereby elected. Being shut out by Jefferson, Aaron Burr decided to run for governor of New York in 1804, and Hamilton campaigned against him, causing his loss.

Taking offense, Burr challenged him to a duel, which took place on July 11, 1804 in Weehawken, New Jersey. Hamilton fired the first shot, which hit a tree branch, but Burr fired the second, which mortally wounded Hamilton, tearing through his liver and diaphragm and lodging in his vertebra. He was rowed back across the Hudson to his friend William Bayard, Jr.'s house. Paralyzed, he lingered in extreme agony for 31 hours. His request for holy communion was honored by the Episcopal Bishop of New York. In a touching scene, he requested that he see his eight children one more time. They filed into his sick room, and he opened his eyes and saw them all there for the last time. He died at two o'clock on July 12, 1804.

His funeral on July 14 caused all businesses to be closed by the New York City fathers. The two-mile procession had so many participants that it took two hours to complete. Governor Morris gave the eulogy at Trinity Church, and Hamilton was then buried in the churchyard.

FUN FACTS

Alexander Hamilton was involved in the very first sex scandal in American politics. He began an affair with 23-year-old Maria Reynolds in the summer of 1791 that resulted in Reynolds and her husband James blackmailing him. In 1792, James Reynolds revealed to then Democratic-Republican Congressman and future President James Monroe that he had evidence of illicit activity by Hamilton as treasury secretary. When confronted with the accusation, Hamilton then refuted it by producing the letters of blackmail over adultery, which disproved his purported conduct as treasury secretary.

CEMETERY

Trinity Church Cemetery Downtown
89 Broadway
Intersection of Wall Street and Broadway
New York, NY 10006
Tel.: 212-602-0800
Hours: Daily, 8:30 AM–4:00 PM

DIRECTIONS TO GRAVE

Upon entering the church grounds, Hamilton's monument is visible on the left. It usually has flowers and flags surrounding it. His wife Eliza is buried beside him with a flat marker.

CLARE BOOTH LUCE
Congresswoman/Ambassador/Author
Born: March 10, 1903
Died: October 9, 1987

In the mid-twentieth century, Claire Booth Luce was a beacon and a trailblazer, paving the way for women to step up onto the same plane as men. An author, politician, U.S. ambassador, and public figure, she presaged what woman are doing today. Married to Henry Luce, the publisher of *Time, Fortune, Sports Illustrated* and *Life* magazines, she con-

Clare Booth Luce's Grave

tributed journalistically to those publications. Also a playwright, she is famous for her 1936 play *The Women*, which had an all-female cast.

Luce was born Ann Clare Booth in New York City on March 10, 1903 to Anna Clare Schneider and William Franklin Booth. Not married, her parents would separate in 1912. Since her father was a travelling salesman, the family moved around quite a bit, with stints in Memphis, Chicago, Nashville and Union City, New Jersey. She attended cathedral schools in Garden City and Tarrytown, New York, graduating first in her class at 16. Her ambitious mother wanted Clare to be an actress, and she had her Broadway debut in a detective comedy called *The Dummy*. In that same year, she became interested in the women's suffrage movement and worked for the National Women's party in Washington, DC and Seneca Falls, New York.

She wed George Tuttle Brokaw, a millionaire heir to a clothing fortune, in 1923 at the age of 20. They had one daughter, Ann Clare. Realizing that Brokaw was an alcoholic, she divorced him on May 20, 1929.

On November 23, 1935, she wed mega-publisher Henry Luce, adopting Clare Booth Luce as her name. The marriage between the two was a difficult one due to his physical awkwardness and lack of humor, which contrasted with Clare's social poise, wit and fertile imagination. It is said that she gave the idea of *Life* magazine to her husband. He encouraged her to write for *Life* and *Time* magazines, but as she grew more famous it became a balancing act to counter the perception of nepotism. During World War II, she published her interviews with General Douglas MacArthur, Chiang Kai-Shek, Jawaharlal Nehru and General Joseph Stillwell.

On January 14, 1944, her only child, Ann Clare was killed in a car accident at the age of just 19. She had been a senior at Stanford University. To assuage Booth's grief, she sought refuge in religion and psychotherapy. She became the most famous convert of Archbishop Fulton J. Sheen and was received in the Catholic Church in 1946.

In 1942, Luce won a Republican seat in the U.S. House of Representatives from Connecticut's 4th Congressional district. Her platform highlighted winning the war and seeking enduring peace, economic security and employment at its conclusion. She was instrumental in the creation of the Atomic Energy Commission. Booth also warned about the rise of international Communism, which could lead to a form of totalitarianism. She did not seek re-election in 1946.

Campaigning vigorously for Dwight Eisenhower's election as president, she was appointed ambassador to Italy and was confirmed by the Senate in March 1953, the first woman to hold such a high diplomatic post. Her admirers in Italy numbered in the millions as she was referred to as "La Signora," the Lady. She forged a close alliance with then Pope Pius XII. She played a vital role in negotiating a peaceful settlement of the Trieste Crisis of 1953–54, a border dispute between Italy and Yugoslavia, averting a possible war.

Surviving an arsenic scare that came from lead paint dust on her bedroom ceiling, she was debilitated mentally and physically, which caused her to resign in December 1956. In April 1959, however, Eisenhower nominated a recovered Luce to become ambassador to Brazil. Although confirmed by the Senate, Luce heeded her husband's advice not to take the post because she would clash with Senator Wayne Morse, who had led the charge against her nomination and who chaired a key Latin American Senate subcommittee.

In 1964 Henry Luce resigned his tenure as editor-in-chief of *Time*. The Luces retired to their winter home in Arizona and planned a permanent retirement in Honolulu, Hawaii, but Henry's death in 1967 came before they could move. After his demise, Clare moved ahead with the plan, completing construction of a gorgeous beach house and joining Hawaiian high society.

However, by 1977, she was living in Washington, DC, which would become her permanent home in her last years. President Ronald Reagan, a friend of the Luces, awarded her the Presidential Medal of Freedom in 1983. She was the first female member of Congress to receive this award.

Clare Luce died of brain cancer on October 9, 1987 at her Watergate apartment in Washington, DC. She is buried in Mepkin Abbey, South Carolina, a plantation that she and Henry once owned and had given to Trappist Monks. She is joined there by her mother, husband and daughter.

FUN FACTS

Luce had a famously acid wit. Examples are: "No good deed goes unpunished," "Widowhood is a fringe benefit of marriage," and "A hospital is no place to be sick." She retained this humor into her old age.

Archbishop Sheen wrote that Luce was one of the most intelligent people he ever encountered. He described her as sharp as a rapier.

CEMETERY

Luce Family Cemetery
Mepkin Abbey Road
Moncks Corner, SC 29461
No telephone number available
Hours: Daily, 9 AM–6 PM

DIRECTIONS TO GRAVE

Upon entering the cemetery gate, you will see a white cross. In front of the cross lie the graves of Clare and her family.

EDWIN STANTON

U.S. Cabinet Secretary
Born: December 19, 1814
Died: December 24, 1869

Edwin McMasters Stanton was secretary of war under President Lincoln, and as such served the nation in a heroic capacity during this turbulent time. He helped to guide the chief executive through many

Edwin Stanton's Grave

thorny situations that came up during the prolonged war that ultimately resulted in victory and preserved our union. At Lincoln's bedside as he lay dying, he took charge of the incredibly chaotic situation and made sure the machinery of government was uninterrupted as he launched the manhunt for Lincoln's assassin.

Edwin Stanton was born to David and Lucy Stanton in Steubenville, Ohio. Early in his youth he became plagued with asthma, which continued the rest of his life to the point of convulsions. The ministrations of Stanton's father, who was a doctor, did little to alleviate the symptoms. Since he was limited in physical activity, he became an avid reader of books and poetry. The family suffered a blow when Edwin's father died suddenly in December 1827 at his residence, and they found themselves destitute. Edwin, at 13, pitched in and worked at the store of a local bookseller.

Edwin attended Kenyon College, but failed to graduate because he lacked the funds to pay for his education. While there he converted to Episcopalian from Presbyterian. He decided to pursue law and studied under the tutelage of David Collier. He was admitted to the bar and began to work at a prominent law firm in Cadiz, Ohio in 1835.

At the age of eighteen he had met and married Mary Ann Lamson. Soon after, he purchased a home in Cadiz, Ohio, where he was to live and where he moved his mother and sisters. Edwin Stanton became very prominent in the community, particularly in the town's anti-slavery society.

Stanton's relationship with Benjamin Tappan, the head of his law firm, expanded when Tappan was elected a U.S. Senator and asked Edwin to oversee the firm, which was based in Steubenville. The Stantons welcomed two children in Steubenville. A baby girl, Lucy, was born in 1840 but died in late 1841 of an unknown illness. A boy, Edwin, survived. Tragedy struck again when Mary Stanton died of bilious fever in March 1844. Her death sent Stanton into a severe depression. He regrouped and focused on his work over the summer. His fame grew and he parted ways with Tappan. Since his practice expanded to Virginia and Pennsylvania, he concluded that Pittsburgh would be a good base of operations and moved there in late 1847.

In 1856, Stanton remarried. He wed Ellen Hutchinson, sixteen years his junior and a descendant of Meriwether Lewis. Through his friendship with Jerimiah Black, who would become James Buchanan's attorney general, he subsequently agreed to represent the nation's interests in California, which his wife loathed. As 1858 drew to a close, Stanton left San Francisco and came back to Pittsburgh, arriving in February 1859. He advised Buchanan on patronage matters and helped Attorney General Black extensively. He also briefly served as attorney general for three months when Black was transferred to secretary of state.

Very soon, as the Civil War got underway, Buchanan's successor, Lincoln, found that he was dealing with a weak secretary of war in Simon Cameron. Calls for his resignation grew. Lincoln and Cameron agreed that he instead be appointed minister to Russia. With the urging of William Seward, Lincoln appointed Edwin Stanton secretary of war, and he was sworn in on January 20, 1862.

He quickly grasped control of the loosely run department and

Signed photo of Stanton, from the private collection of Vincent Gardino

mended relations with Congress. He also worked to create an effective transportation and communication network in the North. Displeased with General McClellan's performance, Lincoln took away his title of general-in-chief and bestowed it upon Stanton, thereby consolidating all military power under Stanton. Ultimately this proved too much to bear for both Lincoln and Stanton, and Lincoln appointed Henry W. Halleck to the position of general-in-chief.

As the war dragged on, Stanton and Lincoln went through a series of generals, which culminated in the selection of Ulysses S. Grant, and then came the battle of Gettysburg. While a Union victory was achieved at Gettysburg, they both felt that Meade had let General Lee escape when he could have been crushed. Their selection was vindicated when Grant vanquished the Confederates at Vicksburg and turned the tide of the war in the Union's favor.

Lincoln was re-elected with Andrew Johnson as his running mate in 1864, largely on the soldiers' vote, which Stanton aided by allowing them to go home to vote. On April 3, 1865, Richmond fell, and the war was essentially over.

Tragedy struck, of course, when Lincoln was assassinated by John Wilkes Booth on April 14. After seeing that William Seward had also been attacked but survived, Stanton made his way to the Peterson House, where Lincoln lay mortally wounded. Stanton took control of the reins of government. He had Andrew Johnson sworn in as president the next morning, had the city locked down and issued a massive manhunt for Booth, which culminated in his capture and death a week later. The remaining conspirators were subsequently captured and later hanged.

Stanton feuded with President Andrew Johnson over the issue of Reconstruction. At one point Johnson removed Stanton as secretary of war in 1867, but the Senate voted overwhelmingly in January 1868 to reinstate him. Johnson once again removed him in February, but Stanton refused to concede his post. Meanwhile, President Johnson was impeached, but survived dismissal by only one vote on May 26. Upon hearing this, Stanton submitted his resignation.

Stanton resumed his law practice amidst failing health. His asthma flared up and impeded his efforts at campaigning in the election of 1868. At Christmas of that year, he could not walk down the stairs and the family celebrated in his room. Meanwhile he waited for newly elected President Grant to reward him for Stanton's election efforts on his behalf.

The new year saw his condition slightly improve, and after Grant's passing over Stanton for a U.S. Supreme Court seat, another became available a day later with a resignation. Grant offered it to Stanton, who accepted. His nomination was confirmed by the Senate on December 20. Unfortunately, on December 23, Stanton complained of pains to his neck and spine. He died at 3 AM on December 24.

His widow nixed a state funeral and opted for a simple one. On December 27, after services at his home, he was born by caisson to Oak Hill Cemetery, where he was interred beside his son James, who had died in infancy a few years earlier.

FUN FACTS

In 1859, Stanton defended in court General Daniel Sickles, who had murdered the son of Francis Scott Key—Phillip Barton Key—in broad daylight at Lafayette Park for having an affair with his wife Teresa. Stanton won Sickles' freedom by claiming temporary insanity. This was the first instance of an insanity plea in American jurisprudence. The jury deliberated a little over an hour to declare Sickles innocent.

After Lincoln died at 7:22 AM on April 15, 1865, Stanton did not actually utter the immortal phrase, "Now he belongs to the ages." What he said actually was, "Now he belongs to the angels."

CEMETERY

Oak Hill Cemetery
3001 R Street
Washington, DC 20007
Tel.: 202-337-2835
Hours:
    Mon.–Fri., 9 AM–4:30 PM
    Saturday, 11 AM–4 PM
    Sunday, 1 PM–4 PM

DIRECTIONS TO GRAVE

Go through the main gate at R Street. Turn right and walk along the path. You will pass the chapel. About 100 feet past the chapel on the left, you will see Stanton's obelisk just off the path.

RAYMOND DONOVAN

U.S. Cabinet Member
Born: August 31, 1930
Died: June 2, 2021

> *The credit belongs to the man who is actually in the arena, whose face is marred by dust and sweat and blood; who strives valiantly; who errs, who comes short again and again, because there is no effort without error and shortcoming; but who does actually strive to do the deeds; who knows the great enthusiasms, the great devotions; who spends himself in a worthy cause; who at the best knows in the end the triumph*

of high achievement, and who at the worst, if he fails, at least fails while daring greatly, so that his place shall never be with those cold and timid souls who neither know victory nor defeat. —Theodore Roosevelt

How does a man who was brought up on charges of grand larceny rate as being a hero in our opinion? Because the charges against Raymond Donovan were ultimately shown to be politically motivated after a jury very quickly determined that his verdict was not guilty. In our research of Raymond Donovan, we were impressed by the way he conducted himself before and after the rendering of this verdict. Our research uncovered, from those who knew Raymond Donovan well, that he was often described as being both an honest and extremely generous man.

Ray Donovan was born in Bayonne, New Jersey, and came from a large family (he had 11 siblings!). As a young man, he considered becoming a Catholic priest. He entered the seminary for the order of the Holy Trinity, where Ray studied in the states of both Alabama and Louisiana. However, he felt at the time that it was necessary that he contribute financially to his family, so he instead took a job with the American Insurance Company. A few years later, he joined the Schiavone Construction Company, where Raymond quickly became a partner.

In 1957 he married Catherine Sblendorio. This union produced two sons and a daughter. They remained married for 63 years until his death. Also in 1957, he co-founded the Fiddler's Elbow Country Club by boldly obtaining a $300,000 bank loan for land that was lying dormant. He shrewdly went to the bank dressed, not in a suit, but in workday clothes sporting a checkered red shirt.

Through hard work and perseverance, Raymond J. Donovan became a wealthy man. Originally Donovan, politically speaking, was a registered Democrat. He supported Adlai Stevenson for president in the 1950s and JFK in 1960. However, as a business executive, he saw the effects of excessive regulation on business. Philosophically, he found himself more and more attuned to the Republican party.

In the 1970s Raymond Donovan met former California Governor Ronald Reagan, and fund-raised for Reagan in his presidential run in 1980. In 1981, Reagan tapped Donovan to be his secretary of

Temporary container of Raymond Donovan's ashes

labor. As such, Donovan worked to reduce regulations on businesses through changing the Occupational Safety and Health Administration (OSHA) enforcement operations. In October 1984 Donovan was indicted on grounds of grand larceny and fraud. This caused Mr. Donovan to leave the Reagan administration in 1985. In a nutshell, the charge against Donovan centered on his construction company trying to defraud the New York City Transit Authority. The evidence against Donovan was so flimsy that some of the jurors, after rendering their verdict of not guilty, commented that the charges were spurious and that the indictment should never have been sought.

Raymond Donovan remained attuned to the political scene and was actively involved in the running of Fiddler's Elbow, ranked as one of the finest country clubs in the United States. He passed away on June 2, 2021 at his home at the age of 90 as the result of congestive heart failure. His cremated ashes will be interred in a columbarium that, as of this writing, is not yet completed. The columbarium is being constructed within The Tower of Remembrance and is projected to be completed sometime in 2024. This property is owned by the Shrine of St. Joseph Church in Sterling, NJ, where Mr. Donovan worshipped and was involved by being a board member and benefactor. Raymond Donovan played a major role in obtaining steel that was part of 9/11's debris for use in the construction of the tower. The bells that hang in the tower were obtained from a seminary in Virginia that was of the order of the Most Holy Trinity. The tower contains over 3,000 names of those who died on that fateful day in 2001.

FUN FACT

Raymond Donovan said shortly after hearing of his not guilty verdict, "What office do I go to to get my reputation back?"

ADDRESS OF FUTURE COLUMBARIUM

The Shrine of St. Joseph
1050 Long Hill Road
Sterling, NJ 07980
Tel.: 908-647-0208
Hours: 7 AM–8 PM

CHAPTER FOUR
Astronauts/Pilots

SALLY RIDE

Astronaut
Born: May 26, 1951
Died: July 23, 2012

Sally K. Ride was the first American woman to fly in space, in addition to being the youngest person to do so (at the time) at age 32.

Sally was born in Encino, California to Dale and Carol Ride, both elders in the Presbyterian church. She grew up in the Van Nuys and Encino areas of California, and went on a yearlong European vacation with her parents where, in Spain, she took up tennis. In 1963, she became a ranked player in her high school and received a tennis scholarship at the Woodlake School, a private and exclusive all-girls school in Los Angeles. She indicated that early that she wanted to become an astrophysicist. She graduated in 1968 and applied to Swarthmore College in Pennsylvania, where she was awarded a full scholarship. There she excelled, and became the Eastern Intercollegiate Singles Champion in tennis.

But she became homesick for California and returned in January 1970. Her aim now was to become a professional tennis player. Sally applied for a transfer as a junior to Stanford. She graduated from there in 1973 with a BS in physics and a BA in English literature. This was followed by an MS in Physics in 1975 and a Doctor of Philosophy in 1978 from Stanford. At Stanford, Sally met tennis professional Billie Jean King, who became a mentor and friend. During her ten-

Sally Ride's Grave
(Courtesy of Susan Lukenbill)

ure at Stanford, Sally decided that she was not up the rigors that are demanded of a tennis pro and gave up pursuing tennis as a career.

In January of 1977, Sally read in the *Stanford Daily* that NASA was recruiting for a new group of astronauts that would include women for the Challenger program. She applied as one of over 8,000 applicants. She went through an endless battery of interviews and medical tests as part of the process, and became one of 35 astronauts in January 1978 as part of NASA Group 8. Six of the group were women. On August 31, NASA announced that the candidates had completed their training and were certified as astronauts.

Ride had several romantic relationships early in her life. The first was with a woman named Molly Tyson. When Ride got into the space program, she dated astronaut trainees, and eventually married fellow astronaut Steve Hawley in 1982.

In April 1982, it was announced that Ride would become the first American woman in space on board the Challenger Space Shuttle, STS-1. It lifted off on June 18, 1983 from the Kennedy Space Center. Part of her job on that mission was to operate the robotic arm to retrieve the Shuttle Pallet satellite, which carried ten experiments to study the formation of metal alloys in space.

She was quickly selected to reprise her performance in space yet again on October 5, 1984 on the Challenger Shuttle STS-41-G. Her prior fight experience allowed her to flawlessly perform complicated repairs involving antennas that were broken or malfunctioning. On these two flights she logged over 343 hours in space. A future third flight was cancelled due to the Challenger disaster in January of 1986.

Ride was appointed to the Rogers Commission, a presidential commission to investigate the disaster. She was the only NASA astronaut on the commission. It was discovered, as may be remembered, that the O-Rings that functioned as a joint seal became stiff at low temperatures, and that had led to the explosion.

In 1987, Ride left NASA to take up a two-year fellowship at Stanford. While there, she divorced her husband and continued a relationship that she had begun with Tam O'Shaughnessy, a former tennis associate.

In 1989, she became a professor of physics at the University of California San Diego and Director of the Cal Space Institute for the university. She turned down offers from President Clinton to become NASA's administrator. In 2001, along with her partner Tam, she founded the Sally Ride Science, a company that created entertaining space programs for elementary and middle school students.

In March 2011, Ride was diagnosed with pancreatic cancer and succumbed to the disease on July 23, 2012, aged 61 years. Her ashes were interred next to her father in Santa Monica. Her obituary publicly announced that O'Shaughnessy had been her life partner, making Ride the first LGBTQ+ astronaut. Her sister Bear confirmed the relationship and said that Ride wanted to keep her personal life private, including her illness and treatments.

FUN FACT

While on her second space mission, Sally carried a white silk scarf that had been worn by Amelia Earhart.

CEMETERY

Woodlawn Cemetery
1847 14th Street
Santa Monica, CA 90404
Tel.: 310-458-8717

DIRECTIONS TO GRAVE

Enter at the 14th Street entrance closest to Michigan Avenue. Continue straight on the road until you reach the end. The grave is in the section directly on the left, about 150 feet from the road and six rows from hedges at the back.

AMELIA EARHART
Aviator/Feminist
Born: July 24, 1897
Died: July 2, 1937

Amelia Earhart was a pioneering aviator and remains an American feminist icon. Her disappearance over 80 years ago still sparks widespread interest as to what happened to her in her last ill-fated flight. She was the first female aviator to fly solo across the Atlantic Ocean

Earhart's signature, from the private collection of Vincent Gardino

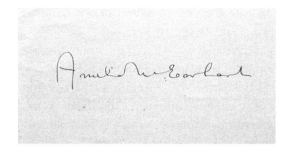

in 1932. (The first individual to successfully fly across the Atlantic was accomplished five years earlier by Charles Lindbergh.) Earhart was also one of the first aviators to promote commercial air travel, and wrote best-selling books about her exploits.

Earhart was born in Atchison, Kansas and was the daughter of Edwin and Amelia Earhart. Little Amelia and her sister, Grace, were brought up in an unconventional manner, as they wore bloomers instead of dresses. Their mother did not want them to be brought up as "nice little girls." Amelia has been described by several biographers as a tomboy. In 1907 her dad's job as a Rockland Railroad claims officer led to a transfer to Des Moines, Iowa, where Amelia was enrolled in public schools. Eventually Amelia and her family moved to Chicago, where she graduated from Hyde Park High School in 1916.

At an early age she showed an interest in flying. At the onset of World War I, she volunteered and received training as a nurse's aide at a military hospital. She was enthralled by stories that were told her by the patients who had flown in combat. In 1920, as her family had moved to California, she persuaded her father to book her on a passenger flight from Long Beach to Los Angeles. When the flight was over, she said, "I knew I had to fly."

Through hard work, persistence and enthusiasm, she completed her flying instructions. Earhart cropped her hair to emulate other female pilots, and in 1922, she purchased a yellow Kinner Airster biplane, which she christened "The Canary." In October of that year she flew her biplane to an altitude of 14,000 feet, setting a world record for a female pilot.

Due to financial hardship, Amelia moved to Medford, Massachusetts. Though she was limited financially in pursuing flying, Earhart became the vice president of the American Aeronautical Society. She promoted flying by writing columns for local newspapers. In April 1928, Amelia was offered the opportunity to fly as a passenger across the Atlantic by Captain Hilton H. Railey, who was an advisor to Admiral Richard Byrd. Though Earhart would not be piloting, she jumped at the chance, and when the flight was completed, she was treated as a heroine. The trip, which was considered potentially hazardous, was sponsored by George Putnam, a publisher and writer, whom Amelia would marry in 1931.

Amelia Earhart's view on marriage was unconventional, particularly for the times. She called her marriage to George Putnam a partnership, with each assuming "dual control." She kept her own name and stated that any moral code regulating faithfulness for both parties was nonexistent.

Earhart earned the nickname "Lady Lindy" after Lindbergh. In 1930, she became active with an organization called the Ninety Nines, which promoted moral support for fledging female pilots.

On the morning of May 20, 1932, the 34-year-old Earhart left Harbour Grace, Newfoundland on her quest to become the first female pilot to fly solo across the Atlantic. This she accomplished in 14 hours and 56 minutes when she landed at Culmore, Northern Ireland.

In 1936 Amelia started planning an around-the-world flight, and with financing provided by Purdue University, had a custom Lockheed Electra twin-engine monoplane fitted with additional fuel tanks. She selected Fred Noonan as her navigator for the proposed flight.

After a bad initial run in which the plane was severely damaged, and which Earhart had luckily escaped unhurt, the plane was successfully repaired. Earhart and Noonan left Miami, Florida on June 1, 1937 with stops in South America, Africa, the Indian sub-continent, and Southeast Asia. They arrived in Lae, New Guinea on June 26. The remaining 7,000 miles would be over the wide Pacific Ocean. On July 2, they departed Lae with a heavily loaded plane (with 1,100 gallons of gasoline), their destination being Howland Island.

At 5 AM on July 2, Earhart reported her altitude was 7,000 feet. The cutter U.S.C.G.C. Itasca was sent close to Howland Island to provide a homing signal for Amelia and her navigator to find Howland more easily without any celestial navigation. The Electra, however, failed to set up any communication with the Itasca. Through a series of misunderstandings or errors, the final approach to Howland Island was not successful. Earhart's 7:58 AM transmission on July 2 said she could not hear the Itasca's signal, but her last transmission at 8:43 AM seemed to indicate that she and Noonan had reached Howland's chartered position, stating "We are running on line north and south."

After one hour past the last transmission from Amelia, the Itasca undertook a search of Howland Island that continued for three days, but to no avail. The U.S. Navy soon joined the search, which continued until July 19th at a cost of over $4,000,000, the largest amount ever spent on a recovery effort up to that time. Amelia's husband also mounted a private search, but again it was futile. Amelia Earhart was declared legally dead on January 5, 1939.

Why Earhart's plane went down and where exactly is the wreckage is an enduring mystery. Titanic Ocean explorer Robert Ballard as recently as 2019 searched the sea floor and found nothing.

The most credible interpretation of her disappearance—from data that has been examined by noted Navy experts such as fuel analysis, radio calls, and other items—strongly point to the fact that the plane went into the water off Howland Island. Her stepson, George Putnam, Jr., has been quoted as saying, "The plane just ran out of gas." Tom D. Crouch, senior curator of the National Air and Space Museum, has said that the Electra rests on the ocean floor approximately 18,000 feet down, and he compares its archeological significance to that of the Titanic. The search for Amelia Earhart continues!

FUN FACTS

When Amelia landed in Culmore, Northern Ireland, completing her solo trans-Atlantic flight, a farm hand approached her and asked, "Have you flown far?" Earhart replied, "from America."

One of her initial endorsements was for Lucky Strike cigarettes. In an era when women's smoking was still frowned upon, *McCall's Magazine* withdrew an offer previously made to Earhart for her to write several articles for the magazine. The $1,500 she would have earned was offered instead for Admiral Richard Byrd's imminent South Pole expedition.

CEMETERY: The Pacific Ocean.

VIRGIL "GUS" GRISSOM
Astronaut
Born: April 3, 1926
Died: January 27, 1967

"Gus" Grissom was one of the leading stars of the inaugural team of astronauts who went into space in the 1960s. He is often overlooked because his first Mercury flight was sandwiched between the first flight into space (by Alan Shepard) and the third, which was the first solo orbital one (by John Glenn). Grissom commanded the first Gemini mission, which completed three orbits with two astronauts on board. He was named commander of the very first Apollo mission, but perished along with his crew in a prelaunch test fire in 1967.

Virgil Evan "Gus" Grissom was born in the small town of Mitchell, Indiana to Dennis and Cecile Grissom. His dad was a signalman for the Baltimore and Ohio Railroad. As a young child, Gus expressed an interest in flying and building model airplanes.

Grissom started attending Mitchell High School in 1940. He wanted to play varsity basketball but was too short, and instead opted for swimming. He graduated in 1944, excelling in mathematics, and spent some time at the local airport in Bedford, Indiana, where a local attorney who owned a small plane took him up on flights and taught him the basics of flying.

Gus met his future wife, Betty, in high school. They married on July 6, 1945, and the union produced two sons, who also had aviation-related careers. Gus valued his home life and was devoted to his family. He shared his favorite pastimes of hunting and fishing with his sons.

World War II began while Gus was in high school, and he was eager to sign up, joining the U.S. Armed Forces in November 1943. Although he was interested in becoming a pilot, he was a clerk prior to his discharge in 1945. He was determined to attend college, and using the GI Bill for partial payment, he enrolled in Perdue University. With him and his wife sacrificing and working menial jobs to pay his tuition, he graduated with a BS in electrical engineering in February 1950.

After graduation he re-enlisted in the military with the U.S. Air Force. In March 1951 Grissom received his wings, and was commissioned a second lieutenant stationed in Presque Isle, Maine. With the ongoing Korean War, Grissom's squad was dispatched to the war zone in February 1952, and he ended up flying over 100 combat missions in six months. On March 11, 1952, he was promoted to first lieutenant and cited for his "superlative airmanship" when he flew cover for a photo reconnaissance mission.

In 1956, he entered the U.S. Air Force Institute of Technology. In May 1957, after attaining the rank of captain, he became a test pilot for the fighter branch.

In 1959, he received an official teletype message instructing him to report to an address in Washington, DC.

Virgil Grissom's Grave

While there he learned of the embryonic space program about to be launched. Extremely interested, and aware of the fierce competition that loomed for the final spots among the 110 test pilots, he entered the competition. On April 13, 1959, he was notified that he was one of the seven Project Mercury astronauts selected, and he was asked to report to Langley Air Force Base in Virginia for astronaut training.

He was selected for the second Mercury flight and made a suborbital flight on his spaceship, Liberty 7. It lasted 15 minutes and 37 seconds on July 21, 1961. After the splashdown, he encountered major problems when the Liberty 7's explosive bolts unexpectedly fired and blew off the escape hatch. This caused water to flood the cabin, and he escaped through the hatch into the ocean, where Gus encountered yet another problem. His space suit began losing buoyancy due to an open-air inlet and he struggled to stay afloat and alive. Fortunately, the recovery helicopter from the *U.S.S. Randolph* came in the nick of time and rescued him. He was later exonerated from having caused any physical or mechanical malfunction of the explosive hatch. The cause has never been explained.

On March 23, 1965, Grissom was the command pilot in the first Gemini flight that carried two astronauts. Along with John Young, he orbited the Earth three times. In a nod to his first capsule sinking, the nickname of the first Gemini spacecraft was Molly Brown. This time the splashdown had no complications.

Gus was then named to head up the first Apollo space mission, leading a crew of three including Ed White (the first man to walk in space) and novice astronaut Roger Chaffee. The official name was Apollo 1.

Problems with the simulator caused Gus to be known as "Gruff Gus" from his complaints about the technical problems they continually ran into. Despite this, NASA pressed on. The problems culminated on February 21, 1967, when the Command Module simulator caught fire and burned up the interior, asphyxiating the three astronauts sealed inside. Just prior to his death Grissom remarked, "How are we going to get to the moon when we can't even talk between two or three buildings?" This was seconds before his shouting "Fire!"— his last word. The fire's ignition source was found to be faulty wiring, which burst into flame in the 100-percent-oxygen pre-launch atmosphere.

On January 31, Gus's funeral and interment ceremonies took place at Arlington Cemetery with President Lyndon Johnson, members of Congress and fellow astronauts in attendance. He is buried beside Roger Chaffee. Edward White was buried at West Point's cemetery.

Gus Grissom accepted the risks inherent in the space program and still adhered to its goals, stating, "If we die, we want people to accept it. We are in a risky business, and we hope that if anything happens to us it will not delay the programs. The conquest of space is worth the risk of life." Prophetic words, to say the least. Betty Grissom later said that perhaps Gus had had a portent of what was to come with Apollo 1. Shortly before returning to Cape Kennedy for the final pre-launch tests, he took a lemon from a tree in his back yard and he explained that he planned to hang it on the spacecraft, although he hung it instead on the simulator where he ultimately died.

The Apollo program moved forward with the launch of Apollo 7, headed by Walter Schirra. The ultimate goal of landing on the moon was attained on July 20, 1969, with Apollo 11. It is ironic that Gus perished because many felt that with his credentials, he was destined to command that first flight to land on the moon. That honor, as we know, went to Neil Armstrong.

FUN FACT

On July 1, 1999, in an expedition funded by the Discovery Channel, Liberty 7 was located at a depth of 16,000 feet (deeper than the Titanic) and retrieved from its watery grave. On July 21, it was returned to Cape Canaveral exactly 38 years after it had sunk.

CEMETERY

Arlington National Cemetery
End of Memorial Avenue, which extends from the Memorial Bridge
  in Arlington, Virginia
Tel.: 877-907-8585
Hours: Daily, 8 AM–5 PM

DIRECTIONS TO GRAVE

Turn right on the Arlington Welcome Center, which is indoors, and immediately turn left on paved walkway. Turn right to stay on paved walkway. Bear left on paved crosswalk. Turn left on Roosevelt Drive Road. Sharp right on Porter Drive Road. Make sharp left on Grave Row intersection. Turn right on Miles Drive Road. Turn left on Grave Row intersection. Grissom is on the right next to his fellow astronaut Roger Chaffee.

ED WHITE
Astronaut
Born: November 14, 1930
Died: January 27, 1967

Edward Higgins White II was an American astronaut who became the first American to walk in space, an accomplishment that had the world take notice of the American space program in the 1960s. He was a member of the crews of Gemini 4 and Apollo 1.

Ed White was born in San Antonio, Texas, the son of Edward Higgins White, Sr. and Mary Rosina White. His father, a West Point graduate, rose to the rank of major general in the U.S. Air Force. Ed expressed an interest in aviation at a very young age when his dad took him for a ride in a North American T-6 Texan trainer.

Ed White's Grave

Due to the nature of his dad's job, the family moved quite a bit. In Dayton, Ohio he attended Oakwood Junior High School, and in Washington, DC he graduated from Western High School in 1948. He wanted to follow in his father's footsteps and attend West Point. Due to the family's constant shuffling, they were not in one place long enough to get a Congressman to recommend him. White took it upon himself to go knock on the doors of the nation's capital with a glowing recommendation from his high school principal. He finally secured an appointment by Congressman Ross Rizley from Oklahoma.

White entered West Point on July 15, 1948. His nickname was "Red" due to the color of his hair. He was active in sports, competing in track, soccer, squash, handball, swimming and golf. He graduated in 1952 with a BS degree and was commissioned a second lieutenant in the Air Force. This was a result of the 1949 agreement where 25 percent of the graduating classes of West Point could volunteer for the Air Force. He received his jet training at James Connolly Air Force Base, Texas, attaining his wings in 1953, and was assigned to Luke Air Force Base in Arizona. That same year, he married his wife Patricia, whom he had met at a West Point football game. They had two children, a son and a daughter.

White was assigned to the 22nd Fighter Squadron at Bitburg Air Force Base in West Germany. There he remained for three and a half years flying North American F-86 and North American F-100 Super Sabre fighters. An article that he read in 1957 prompted White to become interested in being an astronaut. He felt he should increase his chances of being selected. He enrolled in the University of Michigan under Air Force sponsorship to study aeronautical engineering. He was awarded his MS degree in 1959.

Seeing that being a test pilot also improved his chances of selection, he attended the U.S. Air Force Test Pilot School at Edwards Air Force Base in California and graduated in July 1959. As a weightlessness and flight-training captain, he trained future astronauts. One of the men he instructed was John Glenn.

White was one of 11 pilots whom the Air Force submitted as potential candidates for the second group of astronauts. After undergoing medical and psychological examinations at Brooks Air Force Base in San Antonio, he was chosen as one of the nine men in Astronaut Group 2 on September 17, 1962.

He was teamed with astronaut Jim McDivitt as his command pilot for Gemini 4. Early in the planning, EVA (exterior vehicle activity) was planned—this meant a spacewalk. It had been done by a Russian cosmonaut in March 1965, so when Gemini 4 was launched on June 3, 1965, the stage was set for an American walk during the four-day mission. On the third orbit, White stepped outside the capsule and became the first American to walk in space, using an oxygen-propelled gun to maneuver himself. He found the 36-minute walk so exhilarating that he wanted to extend the allotted time and had to be asked to come back in. There was a mechanical problem with the hatch mechanism that made it difficult to open and relatch. McDivitt was able to get it locked by using his glove to push on the gears that controlled the mechanism. This added to the time constraint of the spacewalk. Also, if McDivitt had not been successful, they could not have re-entered the Earth's atmosphere with an unsealed latch.

Upon their return, they were feted as celebrities and presented the NASA Exceptional Service Medal at the White House by President Johnson. Under the usual rotation of the Gemini Program, White would have been in line for his second flight as the command pilot of Gemini 10. Instead, he was selected as the senior pilot of the first manned Apollo flight, Apollo 1, planned for February 21, 1967. For this flight, his fellow crew members were to be Command Pilot Gus Grissom and Pilot Roger Chaffee, who would be embarking on his inaugural space flight.

On January 27, with the crew members mounted atop the Saturn Booster on Launch Pad 34 for a test of the aircraft, a tragic malfunction cost the lives of the three astronauts on board as a fire broke out in the oxygen-filled cabin and very quickly engulfed the capsule in flames. White had attempted to open the inner hatch release handle, and he was found with his hands over his head, trying to open the hatch. The intense heat raised the cabin pressure to the point where the cabin walls ruptured. A spark that jumped from a wire under Gus Grissom's seat was the cause.

Despite pressure by NASA officials to bury White in Arlington alongside Grissom and Chaffee, his widow Patricia stuck to her guns, knowing that his wish was to be buried at West Point. At the funeral, Vice President Humphrey and Lady Bird Johnson represented President Johnson, who was attending the two funerals of Grissom and Chaffee in Arlington. White would later be joined by both his father and son at the West Point Cemetery.

FUN FACT

White was a devout Methodist. On the Gemini 4 mission, he carried three pieces of religious jewelry to take with him on his EVA—a gold cross, a St. Christopher medal, and a Star of David.

CEMETERY

Post Cemetery
329 Washington Road
West Point, NY 10991
Tel.: 845-938-2504
Hours: Daily. 9 AM–4:45 PM

You must go through a security check at the Visitors Center before going to Post Cemetery on the property.

DIRECTIONS TO GRAVE

Upon entering Post Cemetery, turn right and walk down the road until you reach the end of a semi-circle. The path becomes much smaller then.

NEIL ARMSTRONG
Astronaut
Born: August 5, 1930
Died: August 25, 2012

Neil Armstrong became an iconic American hero when he became the first man to land on the moon in 1969, fulfilling President Kennedy's goal to land on the moon before the decade was out.

Armstrong was born near Wapakoneta, Ohio, the son of Viola Louise and Stephen Koenig. His love of flying started at the very young age of approximately five years old when he experienced his first airplane flight in Warren, Ohio with his father.

At the age of 17, in 1947, he began studying aeronautical engineering at Purdue University in Lafayette, Indiana. The U.S. Navy paid his tuition as part of the Holloway Plan (for the training of naval officers). Under the Holloway Plan, the U.S. Navy paid for two years of college, two years of pilot training, one year of service as an aviator and then the completion of his bachelor's degree. He became a midshipman in 1949 and a naval aviator the following year. He saw action in the Korean War as a reconnaissance pilot and ended up flying 79 missions over Korea. His regular commission was terminated on February 25, 1952, and he became an ensign in the U.S. Navy Reserve, resigning his commission eight years later.

After service, he returned to Purdue to complete his B.S. degree in aeronautical engineering in January 1955. While there, he met his future first wife, Janet. The couple had three children together—two boys and a girl.

Upon graduation from Purdue, he became a test pilot at the national Advisory Committee for Aeronautics High Speed Flight Station at Edwards Air Force Base. He was the project pilot on Century Series fighters and flew the North American X-15 seven times. He became an employee of NASA when it was established on October 1, 1958.

On November 1, 1958, the project Mercury took hold, but Armstrong was ineligible as a civilian test pilot, as they wanted only military test pilots. In April 1962, NASA launched the project Gemini astronaut program, which had two astronauts as pilots. This time, selection was open to civilian test pilots. Armstrong applied, was accepted and passed the rigorous physical and psychological tests needed to qualify.

His first spaceflight was as the command pilot for Gemini 8 in March 1966. During this mission with pilot David Scott, he performed the first docking of two spacecrafts. The mission was aborted after Armstrong used some of his re-entry control fuel to stabilize a dangerous roll caused by a stuck thruster. The mission was long, lasting 55 orbits. It successfully landed in the western Pacific Ocean.

After the disastrous loss of life on Apollo 1 on January 27, 1967, NASA regrouped and focused on landing on the moon. Due to normal rotation, Armstrong was selected as commander of Apollo 11 on January 9, 1969, with lunar module pilot Buzz Aldrin and command module pilot Michael Collins.

The Saturn V rocket launched Apollo 11 on July 16, 1969 from the Kennedy Space Center. Despite some concerns about a safe place to land, with Armstrong and Aldrin piloting, they landed the lunar landing module, nicknamed "Eagle."

On July 20, Neil Armstrong stated to Houston, "the Eagle has landed." Michael Collins remained on board the Apollo command module, named Columbia, in lunar orbit. Armstrong stepped onto the lunar surface shortly after uttering the famous words, "That's one small step for man, one giant leap for mankind." An estimated 530 million viewers watched worldwide. After two and a half hours, with Aldrin joining Armstrong on the lunar surface, they planted an American flag, completed gathering moon rocks and left behind a plaque stating that they "came in peace for all mankind." After they re-entered the lunar module, the Eagle successfully launched back up to rendezvous with Columbia, the command and service module. They returned successfully to Earth and were picked up in the Pacific Ocean by the aircraft carrier *U.S.S. Hornet*.

After 18 days in quarantine, they were feted across the nation in a 38-day "Giant Leap" tour. After the flight, Armstrong stated that he didn't want to fly anymore. He completed his M.S. degree in aerospace engineering at USC. Neil then accepted a professorship at the University of Cincinnati because they had a small aerospace department, and he felt they would not resent him coming in with only an

Signed photo, from the private collection of Vincent Gardino

M.S. degree. He then took a heavy teaching load, and remained in that position until 1979.

He then became the spokesperson for the American companies Chrysler and General Time Corporation. He also joined various boards of directors such as United Airlines and Taft Broadcasting.

Neil divorced his first wife in 1994 after 38 years of marriage. He subsequently married his second wife, Carol, after meeting her at a golf tournament in June 1994.

Armstrong had bypass surgery at Mercy Faith General Hospital in Cincinnati on August 7, 2012 to relieve coronary heart disease. Initially doing well, he had complications and died on August 25. Later, his sons sued the hospital for malpractice, stating that nurses had disconnected the wires to his temporary pacemaker, which caused internal bleeding. The doctors allegedly had waited too long to operate after taking him initially to the catheterization lab, and this cost Neil his life. In 2014 the family settled for six million dollars.

Despite calls for a state funeral, his widow opted for a small private one, knowing that that was what Neil would have wanted. On September 14, 2012, his cremated remains were scattered in the Atlantic Ocean from the *U.S.S. Philippine Sea*.

FUN FACTS

Neil Armstrong earned a student flight certificate before his 16th birthday, even before he had a driver's license.

Initially Armstrong used to autograph everything, obliging collectors. However, in 1993 he found out that his signatures were being sold online and many were fake. He then ceased signing any autographs.

In May 2005, Armstrong was involved in a legal dispute with his barber of 20 years, who sold some of his hair to a collector for $3,000. The threatened legal action by Armstrong caused the barber to donate the $3,000 to charity, which assuaged Neil.

FRANCIS GARY POWERS
CIA Pilot
Born: August 17, 1929
Died: August 1, 1977

Francis Gary Powers was a pilot who worked for the CIA. His spy plane was shot down while flying a reconnaissance mission in the Soviet Union. The imbroglio that followed caused the infamous "U-2 incident," which occurred in 1960.

Powers was born in Jenkins, Kentucky, the son of Oliver and Ida Powers. His dad was a coal miner. His family eventually moved to Pound, Virginia, another mining town. Powers' father had hopes for his son to become a physician.

He graduated with a bachelor's degree from Milligan College in Tennessee in 1950, and enlisted in the U.S. Air Force that October. He was commissioned a second lieutenant in December 1952 after completing his advanced training at Williams Air Force Base in Arizona. He was then assigned to the 468th Strategic Fighter Squadron at Turner Air Force Base, Georgia as a Republic F-84 Thunderjet pilot.

In January 1956 he was recruited by the CIA. He was discharged from the Air Force that same year with a rank of captain. He then joined the CIA's U-2 program, whose pilots flew espionage missions at altitudes up to 70,000 feet, supposedly out of reach of Soviet defenses. His plane was equipped with a state-of-the-art camera designed to take high-resolution photos of military installations and other important strategic sites. The primary mission of the U-2 planes was to spy on the Soviet Union. On May 1, 1960, Powers' U2A left a military air base in Pakistan with a dangerous assignment. It was to fly all the way across the Soviet Union, taking a big gamble with a planned route that would go deeper into Russia than ever before and exposing targets that had never previously been photographed.

Francis Gary Powers' Grave

While on the flight, Powers was shot down over Sverdlovsk by one of fourteen S-75 Dvina surface-to-air missiles. The missile struck the tail of the plane and sent it spiraling downward. Powers was unable to activate the camera's self-destruct device before he ejected from the plane. While he parachuted down, he had time to scatter his escape map and partially rid himself of his suicide device. Hitting the ground hard, he was captured and taken to Lubyanka Prison in Moscow.

Initially, the U.S. denied that the plane was anything but a weather plane that had strayed off course. What the CIA failed to realize was that the plane crashed almost fully intact with all the spy camera equipment. Initially, the media depicted Powers as an all-American hero.

In the course of the subsequent trial for espionage, Powers confessed and apologized for violating the Soviet airspace. In the wake of his apology, the media now depicted Powers as a coward. In fact, Powers limited what he told the Soviets about his CIA contacts, particularly about the important element of the height he was travelling, which he said was the maximum altitude of 68,000 feet. In fact it was 70,000 feet. It was believed by U.S. aerial engineers that at 70,000 feet, planes would go undetected on radar. At his trial, he was convicted of espionage and sentenced to 10 years of confinement. Three were to be in a prison and the subsequent seven in a labor camp.

He was first imprisoned in Vladimir Prison, 150 miles east of Moscow, from September 9, 1960 to February 8, 1962. Despite opposition from the chief of the CIA counterintelligence for a prisoner swap, President Kennedy signed off on an exchange. In the place of KGB Colonel William Fisher, known as Rudolf Abel, Powers and the U.S. student Frederic Pryor were set free on February 10, 1962. Powers had been confined one year, nine months and 10 days.

Initially Powers received a cool reception upon his return to the States and was criticized for not activating the plane's self-destruct charge for the camera, film, and related classified parts. He was even criticized for not taking the suicide pill that he had on his person. After being debriefed by the CIA, Director John McCone exonerated Powers and stated that he had fulfilled his employment and his instructions in connection with the mission. On March 6, 1962 he appeared before the Senate Armed Services Committee. At the end of the hearing he was given sustained applause by the senators. Senator Prescott Bush declared, "I am satisfied he has conducted himself in an exemplarily fashion and in accordance with the highest tradition of service to one's country." Senator Barry Goldwater, who was a pilot himself, sent him a handwritten note that stated, "You did a good job for your country."

He had also been scheduled for a White House meeting with President Kennedy later that day, but it had been cancelled due to pressure by Attorney General Robert Kennedy, who felt the hearing would not go well. It had so far gone the opposite way for Powers, but unfortunately the meeting had already been cancelled and he never met with JFK.

The whole ordeal took a toll on his marriage to his first wife, Barbara, who became an alcoholic and had numerous affairs while Powers was in captivity. Upon his release she constantly threw tantrums, cursing him in public and overdosing on pills. He sued his wife for divorce on August 10, 1962. He then started a relationship with Claudia Edwards "Sue" Downey, whom he had met working briefly at CIA headquarters. She had a daughter from a previous marriage. They married on October 26, 1963 and had a son, Francis Gary Powers, Jr. His wife worked hard to preserve his legacy after his untimely death.

Powers worked for Lockheed as a test pilot from 1962 to 1970, though the CIA paid his salary. The publication of his book, *Operation Overflight*, had ruffled some feathers at Lockheed and resulted in his termination. Powers then became a reporting airline pilot for Buckley Broadcasting's radio station KGIL in the San Fernando Valley. After a few years, he became a helicopter traffic reporter for KNBC-TV, the NBC O&O in Los Angeles.

On August 1, 1977, he and his cameraman, George Spears, were killed when their Bell JetRanger helicopter crashed after they were returning from filming brush fires in Santa Barbara County. According to Powers' son, an aviation mechanic repaired a faulty fuel gauge without informing Powers, who subsequently misread it. In a final act of bravery, it is surmised that he noticed children playing in the area and diverted the helicopter elsewhere to avoid landing on them. He might have landed safely if not for the last-second deviation, which compromised his autorotative descent. He was interred at Arlington National Cemetery as an Air Force veteran.

In death, Powers was certainly vindicated with numerous subsequent military honors including the Prisoner of War Medal, Distinguished Flying Cross, National Defense Medal, the Silver Star Medal and the highly coveted CIA Director's Medal. Despite the secretive nature of the CIA, in a speech in March 1964, former Director Allen Dulles stated, "He performed his duty in a very dangerous mission, and he performed it well, and I think I know more about that than some of his detractors and critics know, and I am glad to say that to him tonight." What did Dulles know? We might never know because quite a bit of the incident remains classified to this day.

FUN FACT

While in captivity, Powers learned carpet weaving from his cellmate, a Latvian political prisoner, to pass the time.

CEMETERY

Arlington National Cemetery
End of Memorial Avenue, which extends from the Memorial Bridge in Arlington, Virginia
Tel.: 877-907-8585
Hours: Daily, 8 AM–5 PM

DIRECTIONS TO GRAVE

Turn right on at the Arlington Welcome Center, which is indoors, and immediately turn left on paved walkway. Turn right to stay on paved walkway. Bear left on paved crosswalk. Turn Left on Roosevelt Drive Road. Turn right on Memorial Drive Road and make a sharp left on Wilson Drive Road to stay on Memorial Drive Road. Turn right on Grave Row intersection to stay on Memorial Drive Road. Bear right on Porter Drive Road. Make sharp left on McPherson Drive Road. Turn left on McKinley Drive Road. Turn right on Grave Row intersection. Powers will be on the right.

CHAPTER FIVE
Sports

BABE RUTH
Hall of Fame Baseball Player
Born: February 6, 1895
Died: August 16, 1948

Babe Ruth, whose career spanned 22 seasons in Major League baseball, is regarded as one of the great heroes in sports culture. Many think he was the greatest baseball player of all time. His outstanding performance in the early 1920s is credited with saving the game of baseball after it was severely tarnished with the Black Sox scandal of 1919. He injected excitement into the game with his home-run production, which until his arrival on the scene had been minimal. He became the first player to hit 60 home runs in one season. His colorful lifestyle off the field also added to his legend.

George Herman Ruth, Jr. was born in Baltimore to Katherine and George Ruth, both of German ancestry. Ruth was sent at the age of seven to St. Mary's Industrial School for boys because he was incorrigible. He spent the next 12 years at St. Mary's. It was there that Ruth took up baseball, which was encouraged by the Prefect of Discipline, Brother Matthias Boutlier. Brother Matthias became a mentor to Ruth in both baseball and his personal life. He lavished attention on Ruth and curbed his wild instincts. The school instilled a Catholic culture in Ruth, and later, he often attended mass after a night of carousing. He never forgot St. Mary's and was generous to them when he became famous and rich, helping to fundraise and donating money.

In 1914, Ruth signed a contract to play for the minor league Baltimore Orioles and earned a monthly salary of $100. It was there that he acquired his nickname "Babe" due to his lack of knowledge of proper etiquette. His first professional appearance came on March 7, 1914 when playing shortstop, and he pitched the last two innings of a 15–9 victory.

He began to acquire a reputation as a great pitcher and home-run hitter during the dead ball era, which marked a period of time approximately from 1900 to 1920. It is called the "dead ball era" due to the propensity of low scoring games. The Orioles encountered severe financial difficulties at this time, which forced the owner to sell Ruth to the Boston Red Sox. He won his first game as a starting pitcher on July 11, 1914 against the Cleveland Naps.

Babe Ruth's Grave

Ruth's tenure at Boston highlighted his pitching abilities and home-run hitting prowess. Not happy with playing every four days as a pitcher, Ruth wanted to play every day at another position. GM Ed Barrow began using him at first base and the outfield. Boston owner Harry Frazee, having financial problems as a Broadway empresario, decided to sell Babe Ruth to the New York Yankees despite the fact the Boston Red Sox had won the 1918 title. On December 26, 1919, he sold Ruth's contract to the New York Yankees for $100,000, the largest sum paid for a ballplayer up to that time.

Colonel Jacob Ruppert and his co-owner, Colonel Tillinghast Huston, of the Yankees, embarked on a spree to acquire additional ball players in addition to constructing a new stadium to showcase their team. In 1921, Ruth hit 59 home runs. When the new stadium, dubbed "the House That Ruth Built," opened in 1923, he was coming into his prime. And the right-field fence was closer then, aiding left-handed hitters such as Ruth. That year they christened the new stadium, defeating the

NY Giants in the World Series, with Ruth having an outstanding season batting .393. The 1920s decade with Ruth as the leader had the New York Yankees dubbed as "Murderers Row." In 1927, he hit his record 60 home runs. He became the highest-paid baseball player ever at the time, earning $80,000 annually in 1930.

In the '30s, Ruth initially still retained his hitting skills. He hit the first home run in All-Star game history on July 6, 1933 at Comiskey Park in Chicago, a two-run blast. By 1934, however, his skills had diminished considerably due to years of high living. In 1935 Colonel Ruppert sold Ruth to the Boston Braves. Braves owner Judge Emil Fuchs had hinted at the possibility of making Ruth manager over time, giving him the title of vice president. As the season unfolded, Ruth played poorly, and realized that Fuchs had deceived him as it became evident that Fuchs had no intention of ever making him a manager. Ruth's final game was in Philadelphia, and on June 2, 1935, Ruth called it a day. His career stats were impressive: lifetime .342 batting average, 714 career home runs, and a 2.28 ERA.

Retirement saw Ruth frustrated in his quest to become a manager. He was a first-base coach with the Brooklyn Dodgers in 1938. After one year he left after failing to get along with new manager Leo Durocher. He whittled away his time playing golf in Rye, New York, participating in many charity tournaments. Ruth reconciled with teammate Lou Gehrig, with whom he had not spoken for years, on Lou Gehrig Appreciation Day, July 4, 1939. Gehrig was honored on that day after being diagnosed with ALS. During World War II, Ruth made many personal appearances to aid the war effort, including his last appearance as a player in Yankee uniform in a 1943 exhibition for the Army-Navy Relief Fund.

Ruth was married twice. He met his first wife, Helen Woodford, in Boston and they married as teenagers in 1914. They separated in 1925 because of his repeated infidelities. They last appeared as a couple during the 1926 World series. In 1929, Helen died in a house fire. Three months later, he married actress and model Claire Merritt Hodgson, who unlike Helen was worldly and educated. She put structure into Ruth's life. Through a previous marriage she had a daughter, Julia, whom Ruth adopted.

In 1946 it was found that Ruth had an inoperable tumor in his brain at the base of his skull. In the ensuing two years, he underwent experimental radiation and chemotherapy treatments. On June 13, 1948, he visited Yankee Stadium for the last time for the 25th anniversary of "the House That Ruth Built." He was in bad shape, having lost much weight, and he had difficulty walking, using his bat as a cane. A photograph of him from behind standing at home plate won the Pulitzer Prize for photographer Nat Fein that year. Ruth faded fast and expired at Memorial Hospital on August 16, 1948, not knowing at the end that he had cancer. He lay in state in the rotunda of Yankee Stadium, where 77,000 people filed past his open casket for two days. His funeral at St. Patrick's Cathedral was conducted by Cardinal Spellman with 75,000 waiting outside. Burial followed at Gate of Heaven Cemetery in Hawthorne, New York.

FUN FACTS

The curse of the Bambino carried over even into 2008 when an excavation at Yankee Stadium turned up a dusty Boston Red Sox uniform embedded in 2 feet of concrete. A construction worker had hoped it would jinx the Yankees in their new stadium. The plot was foiled!

In the 1926 World Series, Babe Ruth made a promise to Johnny Sylvester, a hospitalized 11-year-old boy, to hit a home run on his behalf. He did so, and it helped the boy emotionally and helped him to recuperate.

CEMETERY

Gate of Heaven Cemetery
10 West Stevens Avenue
Hawthorne, NY 10532
Tel.: 914-769-3672
Hours: 9 AM–4:30 PM

DIRECTIONS TO GRAVE

After entering at 10 West Stevens Avenue, turn right and follow the main road up the hill. Turn right immediately before Section 25 and Babe's gravesite is 50 yards on the left.

JIM THORPE
Olympic Athlete/Gold Medalist
Born: May 22, 1887
Died: March 23, 1953

Jim Thorpe is generally regarded as the greatest Olympic athlete of all time. In a poll conducted by ABC Sports in 2000, he was voted the greatest athlete of the 20th century. He won two gold medals in the 1912 Olympic games, one in the classic pentathlon, the other in the decathlon. The medals were taken away from him when it was discovered the next year that he had played two seasons of semi-pro baseball while in college. In 1983, the International Olympic Committee (IOC) restored his Olympic medals with replicas. The decision to reverse the 1913 ruling to disqualify Thorpe was based on a violation of IOC rules, as it came outside the required 30 days of his winning the medals.

James Francis Thorpe was born in Oklahoma. Jim's parents were of racially mixed ancestry. His father, Hiram, was Irish, while his mother, Charlotte, was a Potawatomi Indian. Thorpe was raised as a member of the Sac and Fox tribe. Both parents were Roman Catholic and Thorpe remained so the rest of his life.

Jim attended the Sac and Fox Indian School in Stroud, Oklahoma, but left after the death of his younger brother. Hiram then enrolled him in the Haskel Institute, an Indian boarding school. After the death of his mother, he left yet again and worked at a horse ranch.

In 1904, Thorpe returned to his dad and enrolled in the Carlisle Indian Industrial School. He began his athletic career there in 1907, competing in track and field. He would also add football, baseball and lacrosse to his repertoire.

The football coach of Carlisle, Pop Warner, decided to put him on the team. In 1911, he received national recognition for his ability as a running back, defensive back, place kicker, and punter. He scored all his team's field goals in an upset of top-ranked Harvard in the early days of the NCAA. Carlisle finished the season 11–1 and won the national collegiate championship as a result of Thorpe's 25 TDs and 198 points that season. He was named First Team All American in 1911 and 1912. Football always remained his favorite sport.

He began training for the Olympic games in Stockholm, Swe-

Jim Thorpe's Grave

den in 1912. At the games, as we know, he won two gold medals. In addition, he won two challenge prizes in the form of trophies, one of which was donated by King Gustav of Sweden and the other by Czar Nicholas II of Russia. He won eight of the fifteen individual events of the pentathlon and decathlon, accumulating 8,413 points.

In January of 1913, the *Worcester Telegram* reported that Thorpe had played professional baseball before the Olympics. College players regularly played to earn extra money, but they used aliases, which Thorpe did not. First, the Amateur Athletic Union withdrew his amateur status, and then the International Olympic Committee stripped him of all his medals.

In 1913, Thorpe started his professional baseball career, which included stints with the New York Giants, Cincinnati Reds, Chicago

Cubs and Boston Braves. He compiled a .252 lifetime batting average over 289 games.

He also played professional football with the Canton Bull Dogs, one of 14 teams to form the American Professional Football Association, the precursor of the National Football League. He became the first president of what was to be later known as the NFL. He retired from professional football at the age of 41 and was later inducted into the NFL Hall of Fame as a charter member in 1963.

Jim was married three times and had eight children. His last wife, Patricia, became his manager and was with him when he died.

Thorpe found it hard to hold onto a non-sports-related job to provide for his family. At various times he worked as ditch digger, security guard, bouncer, doorman and an extra in movies. He was also briefly in the Merchant Marine during World War II. He was memorialized in the 1951 Warner Brothers film *Jim Thorpe All American*, starring Burt Lancaster.

Toward the end of his life, he was broke and had to rely on charity to get by. On March 28, 1953, at the age of 65, he suffered a fatal heart attack after dining with his wife. She then made the controversial move of having two Pennsylvania towns, Mauch Chuck and East Mauch Chuck, change their names to combine as "Jim Thorpe" in return for her receipt of monetary payment. Part of the deal was the towns' buying his body and erecting a tomb with two statues of him in athletic poses. His body resides there to this day.

The red granite mausoleum rests on mounds of soil from Thorpe's native Oklahoma, the New York Polo Grounds (home of the New York Giants baseball team) and from Stockholm's Olympic stadium.

FUN FACTS

In addition to his athletic prowess at the Carlisle School, Jim was an accomplished ballroom dancer. He won the intercollegiate ballroom dancing competition in 1912.

In the Olympic decathlon, where he placed in the top four in all ten events, he competed in a pair of mismatched shoes because someone had stolen his shoes before he was due to compete in the event. One of the mismatched shoes, as it turned out, he found in a trash can!

BURIAL SITE

The Jim Thorpe Memorial is located along North Street (Rt. 903) in Jim Thorpe, Pennsylvania, part of Carbon County.

JOE DiMAGGIO
Hall of Fame Baseball Player
Born: November 25, 1914
Died: March 8, 1999

Joseph Paul DiMaggio is considered by many to be one of the greatest baseball ballplayers who ever played the game. He was a hero to many Americans, particularly those of Italian descent, who followed his exploits, especially his incredible feat of getting at least one hit in 56 consecutive games. It is a record that still stands today.

He was born to Sicilian immigrants Giuseppe and Rosalie DiMaggio. His father was a fisherman who immigrated to the United States in the hopes of earning a better living. He tried to recruit Joe into fish-

Joe DiMaggio's Grave

ing, but Joe rebelled and turned to baseball. After dropping out of Galileo High School in San Francisco, California, he worked odd jobs. By 1931 he was playing semi-pro ball. In 1932 he made his professional debut as a short stop in the Pacific Coast League with the San Francisco Seals, thus making the jump from playground to the pros in less than two years. In a harbinger of things to come, he hit safely in 61 games in his rookie year of 1933. The New York Yankees bought him from the Seals in 1935.

His storied major-league career began on May 3, 1936, when he batted ahead of Lou Gehrig in the lineup. DiMaggio, with an explosive bat, speed and enthusiasm, helped propel the Yankees to winning the World Series in four consecutive seasons from 1936 to 1939. During his thirteen seasons with the Yankees, he led them to nine World Series championships. Nicknamed the "Yankee Clipper" because of his range and speed, he was also a superstar in batting, excelling in batting average, homers and RBIs. The highlight of his career was his aforementioned 56-game hitting streak in the summer of 1941, with the nation riveted to his daily exploits. The closest anyone has come to breaking this streak was Pete Rose of the Cincinnati Reds, who notched hitting safely in 44 consecutive games. Many consider Joe's an unbreakable record and a statistical impossibility that will stand for eternity.

Before retiring, Joe became the first ballplayer to break the $100,000 barrier in terms of earnings. In 1951, wanting to go out on top, he retired from the game. He had amassed many achievements, such as being named a 13-time All Star, nine times World Series champion, three times AL MVP, two times AL batting champion, two times home run leader and two times RBI leader. He was inducted into the Baseball Hall of Fame in 1955.

His personal life was marked by two marriages, the second of which was the most famous. He married actress Dorothy Arnold in 1939, and that union produced a son, Joe Jr. His second appearance at the altar came in 1954, when he tied the knot to glamorous movie star Marilyn Monroe. Their tempestuous union lasted only nine months, with DiMaggio's jealousy and physical abuse cited as reasons for Marilyn's divorce. The straw that broke the camel's back, enraging DiMaggio, was the infamous skirt-blowing scene above subway grates in *The Seven Year Itch*.

They reentered each other lives, however, as Marilyn's marriage to Arthur Miller was ending. In 1962, DiMaggio secured her release from the Payne Whitney Psychiatric Clinic in Manhattan. It was assumed they would remarry, but Marilyn died tragically in August 1962. Devastated, Joe claimed her body and arranged her funeral, barring Hollywood's elite and the Kennedy family. For twenty years thereafter, he sent a half dozen roses to her crypt three times a week.

He returned to baseball briefly when the Athletics relocated to Oakland in 1968. He served as an honorary vice president in 1968 and 1969 as well as a coach in the first of those two seasons.

Joe became a commercial spokesman in the '70s, '80s and '90s with Mr. Coffee and the Bowery Savings Bank. Retiring to Florida, he also became a spokesman for Florida's Cross Keys Village, an active retirement village. He was responsible for raising four million dollars for the Joe DiMaggio Children's Hospital in Hollywood, Florida, which opened its doors in 1992.

A heavy smoker for much of his life, Joe was diagnosed with lung cancer in 1998 and succumbed to the disease at his home in Hollywood, Florida on March 8, 1999. His funeral was held on March 11 at Saints Peter and Paul Church in San Francisco, the site of his first marriage to Dorothy Arnold. He was entombed at Holy Cross Cemetery three months later in Colma, California.

FUN FACTS

Joe's two brothers were also major league players. Vince was with the Phillies, Reds, Giants and Pirates, and Dominick was with the Boston Red Sox.

Joe could be at times severe and belligerent. He insisted on being introduced as "the Greatest Living Ballplayer" at major events. When comedian Billy Crystal failed to introduce him as such at Old Timers Day at Yankee Stadium, he punched Crystal in the stomach.

CEMETERY

Holy Cross Cemetery
1500 Mission Hill Road
Colma, CA 94014
Tel.: 650-756-2060
Hours: 8 AM to Sunset

DIRECTIONS TO GRAVE

DiMaggio's burial vault is located in Section I. Enter cemetery on Old Mission Hill Road. You will see the office on your left. Proceed down that road, keeping left and passing area D. On your right will be a circular path going around the receiving chapel on the right. Keeping left, follow it around until the road is straight. On the left will be section I. About 100 feet along the road and slightly up an incline, you will spot DiMaggio's black vault surrounded by baseball memorabilia.

JACKIE ROBINSON
Hall of Fame Baseball Player/Civil Rights Activist
Born: January 31, 1919
Died: October 24, 1972

Jack Roosevelt Robinson broke the major league baseball color barrier and played a huge role in the civil rights movement in the United States in the mid-20th century.

Jackie was born to a family of sharecroppers in Cairo, Georgia, the youngest of five children of Mallie and Jerry Robinson. After his father had left the family, in 1920 they moved west to Pasadena, California. At John Muir Tech High School, Jackie played baseball, basketball, track and tennis at the varsity level. In 1937, he attended Pasadena Junior College, where he continued his athletic interests.

While at PJC, he was arrested on January 25, 1938 because he vocally disputed the detention of a black friend by the police. He was given a two-year suspended sentence, but the incident gave him a reputation for combativeness against racial prejudice. Toward the end of his enrollment at PJC, his brother Frank, to whom he was very close, was killed in a motorcycle accident. This prompted Jackie to pursue his athletic career at UCLA to be closer to his brother's family in the spring of 1939. At UCLA, he became the school's first person to win varsity letters in basketball, football, baseball and track.

While at UCLA, he met his future wife Rachel while he was a senior and she was a freshman. She had been familiar with his athletic exploits at PJC. Their subsequent union in 1946 produced three children—two boys and a girl.

Jackie Robinson's Grave

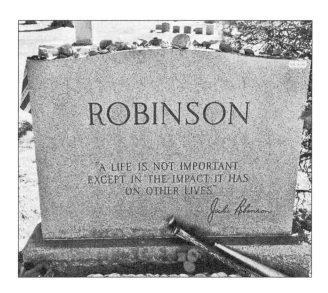

Jackie left UCLA just shy of graduation in 1940 to become assistant athletic director at the government's National Youth Administration. The government ceased its operations just as the Japanese bombed Pearl Harbor in 1941. The following year, Jackie was drafted and assigned to a segregated army calvary unit at Fort Riley, Kansas. Despite prejudice from the army brass, with the help of Joe Louis, the boxing champion also stationed there, he was accepted into officer candidate school. He was commissioned a second lieutenant in January 1943 and reassigned to Fort Hood, Texas. Robinson was court martialed in 1944 when he was accused of making racial accusations against a military police officer, for which he was acquitted by an all-white panel. He received an honorable discharge that November.

Upon his discharge, he wrote a letter to the Kansas City Monarchs in the Negro Leagues, asking for a tryout. While waiting, however, he became the athletic director at Samuel Huston College in Austin, Texas. The Monarchs subsequently responded, offering him a contract to play for $400 per month in 1945.

While playing for the Monarchs, Jackie pursued major league teams including the Boston Red Sox and the Brooklyn Dodgers. Branch Rickey, the president and general manager of the Dodgers, was looking for black players to play on their farm team in Montreal.

In a contentious face-to-face four-hour meeting, he extracted a promise from Robinson to "turn the other cheek" with the inevitable racial animus that would be directed toward him if he played for them. He reluctantly agreed and was signed to a contract at $600 per month.

In 1946, he arrived for spring training at Daytona Beach, Florida, and immediately experienced racial discrimination, starting with segregated hotel facilities. Rickey persevered in getting Robinson needed exposure and some spring training games in Daytona Beach. When the season ended, Jackie led the International League with a .349 batting average and was named MVP. His presence also substantially increased attendance in the league.

The following season saw Jackie break the major-league color barrier. His promotion in 1947 was generally well-received, but there was some dissension on the Dodger team that was quelled by manager Leo Durocher. He said emphatically that he would play Jackie. The St. Louis Cardinals threatened to strike. National League President Ford Frick said he would suspend all players who threatened to do so, and the threat was eliminated. Prominent players such as teammates Pee Wee Reese, Hank Greenberg and Larry Doby also lent Robinson moral support. Rising above the stress of his first season, he won the very first MLB Rookie of the Year Award.

His subsequent major-league career was stellar. His lifetime batting average of .313 over nine seasons included being named MVP in 1949 and winning a world championship in 1955, defeating Dodger archrivals the New York Yankees. The Dodgers traded Jackie to the New York Giants in 1956, but the trade never materialized because he had agreed to quit baseball. Instead, he became the first black vice president of a major corporation, Chock Full O' Nuts.

In his first year of eligibility, 1962, Jackie was enshrined in the Baseball Hall of Fame on the first ballot. In 1965, he continued his trailblazing by becoming an analyst on ABC TV's major league baseball game of the week, the first black person to do so.

During the 1960s, Jackie was plagued by heart problems as well as diabetes, which rendered him almost blind by the end of the decade. In 1968, he suffered his first heart attack. This was followed by a fatal one on October 24, 1972 at his home in North Stamford, Connecticut. Following his funeral at Riverside Church in Manhattan, he was interred in Cypress Hills Cemetery in Brooklyn next to his son. Following his death, he was posthumously awarded the Presidential Medal of Freedom in 1984 and the Congressional Gold Medal in 2005.

FUN FACTS

Jackie's middle name, Roosevelt, was in honor of Theodore Roosevelt, who had died a few weeks before his birth.

Jackie's widow, Rachel, as of this writing has reached her 101st birthday. She founded the Jackie Robinson Foundation, which provides scholarship funds for college students, and where she is still an officer to this day. In 2022 the foundation opened the Jackie Robinson Museum in lower Manhattan. It highlights her husband's legacy as well as focusing on the civil rights movement, the city's first museum to do so.

CEMETERY

Cypress Hills Cemetery
833 Jamaica Avenue
Brooklyn, New York 11208
Tel.: 718-235-6280
Hours: Mon.–Sat., 8 AM–4 PM; closed on Sundays

DIRECTIONS TO GRAVE

Enter the cemetery at 833 Jamaica Avenue, turn right and drive past the office. After the road's left-hand bend, make the first right, and then take the next left. Drive up the hill to Memorial Abbey, and Jackie is buried across the drive from the abbey, next to the Slater Tomb.

JESSE OWENS
Olympic Athlete/Gold Medalist
Born: September 12, 1913
Died: March 31, 1980

James Cleveland Owens was an Olympic track and field athlete who won four gold medals in the 1936 Olympic games in Berlin, specializing in sprints and the long jump. During his lifetime, he was recognized as "perhaps the greatest and most famous athlete in track and field history."

Jesse, a grandson of a slave, was born to sharecropper Henry Owens and his wife Mary Emma in Oakville, Alabama. He was the youngest of ten children. At nine years of age, his family moved to Cleveland, Ohio for more employment opportunities.

As a young man he took menial jobs, but at this time Owens had a passion for running. Jesse attributed his athletic success to Charles Riley, his track and field coach at Fairmont Junior High School. Since Jesse worked after school, Riley allowed him to practice before school opened. It was there that he met his future wife, Minnie Ruth. They subsequently married in Cleveland in 1935 and had three daughters.

He came to national attention in 1933 when he was a student at East Technical High School in Cleveland. He equaled the world's record for the 100-yard dash at the National High School Championship in Chicago.

Owens attended Ohio State University after his dad found employment so their large family could subsist. He encountered racial discrimination and had to live off campus and eat at "blacks only" restaurants. During his time there, however, he won an amazing record eight individual championships. He did not receive an athletic scholarship, so he had to continue to work part-time to pay for school.

On May 25, 1935, in the span of 45 minutes, he achieved track and field immortality by establishing three world records at the Big Ten track meet in Ann Arbor, Michigan. He set three world records in the long jump, 220-yard dash and 220-yard low hurdles. He also tied the world record for the 100-yard dash. In 2005, a University of Central Florida professor deemed that this was, in one day, the most impressive athletic achievement since 1858. It has never been equaled.

Despite backlash from the NAACP and the American Olympic Committee, Owens decided to compete in the 1936 games that were being held in Berlin, Germany, despite the German leader, Adolph Hitler, promulgating a racist Aryan regime. Arriving at the new Olympic stadium, Owens was met with a throng of fans shouting his name. From August 4th to 9th, he won four gold medals in the 100-meter dash, the long jump, the 200-meter sprint and the 4×100-meter

Jesse Owens' Grave

relay. Hitler was initially faulted for not acknowledging Owens' victories, but Owens contradicted this by stating that he had passed Hitler's box on his way to a broadcast interview, Hitler had waved at him and he had waved back. Hitler also sent him a signed commemorative photo even though his victories contradicted the Aryan race theory of superiority. Owens complained instead that President Roosevelt never acknowledged his performance, and after a victorious ticker-tape parade in New York City by Mayor Fiorello LaGuardia, never invited him to the White House to congratulate him. He subsequently joined the Republican Party and supported Alf Landon for president in the 1936 election.

On his return home, despite having won four Olympic gold medals, he faced difficulty finding employment, and ended up working as a gas station attendant, playground janitor and manager of a dry-cleaning store. He also raced against amateurs and horses for cash. Finally, a friend brought him to Detroit to work for the Ford Motor Company as assistant personnel director in 1942. He rose to director and worked there till 1946. In the late 1940s, he and his family moved to Chicago, where they lived until 1972, when he retired to Tucson.

In 1946 he had a brief foray into ownership in a new negro baseball league, the West Coast Negro Baseball League, with the Portland, Oregon franchise "the Rosebuds." Sadly, the league folded after only four months. He lamented that at that time there weren't any endorsements for black athletes, and he was again forced to take menial jobs to support his family. His comeback was aided by President Dwight Eisenhower, who made him a good-will ambassador to India, Malaysia and the Philippines, extolling the cause of American freedom and economic opportunity for the developing world. He would continue this into the '60s and '70s. He lost his additional patronage job with the Illinois Youth Commission in 1960, but forged ahead with product endorsements for Quaker Oats, Johnson & Johnson and Sears. He and his wife retired to Tucson, Arizona in 1972.

He initially refused to support the black-power salute of sprinters John Carlos and Tommie Smith at the 1968 Olympic games. He later reversed his opinion in 1972 in his book *I Have Changed*. In 1972, he travelled to the Olympic games as a guest of the West German government and met Chancellor Willy Brandt and boxing star Max Schmeling. A few months before his death, he unsuccessfully tried to convince President Carter to withdraw his demand that the U.S. not compete in the 1980 Moscow Olympics due to their invasion of Afghanistan. He argued that the Olympic ideal was above politics.

He was diagnosed with an aggressive form of lung cancer that was drug resistant in December 1979. On March 31, 1980, he succumbed to the disease with his family at his side. He was interred at the Oakwoods Cemetery, located in Chicago next to the Lake of Memories.

FUN FACTS

As a child, Owens was nicknamed "JC" when a teacher asked him his name for a listing in the school roll book, and with his strong Southern accent she understood him to say "JC" instead of "Jesse." He was known by that for the rest of his life.

At Ohio State University, he had another unique nickname due to his speed. It was "the Buckeye Bullet."

CEMETERY

Oakwoods Cemetery
1035 East 67th Street
Chicago, Illinois 60637
Tel.: 773-288-3800
Hours: 8:30 AM–5 PM

DIRECTIONS TO GRAVE

Enter the cemetery at the 1035 East 67th Street entrance, turn right, then bear left at the "Y" onto Memorial Drive. After the lake, turn right and Jesse's grave will be on the right.

HANK AARON
Hall of Fame Baseball Player
Born: February 5, 1934
Died: January 22, 2021

Hank Aaron is one of the greatest ballplayers that the game of baseball has ever produced. In the twilight of his career, he shattered a beloved baseball record no one ever thought would be broken. He did it with a grace and dignity that personified his professional and personal life despite having to contend with extensive racist threats. He is the owner of a litany of baseball records, many of them for power-hitting.

Henry Louis Aaron was born in Mobile, Alabama to Herbert and Estella Aaron. As a child, he grew up poor and had to use bottle caps to hit with sticks as he couldn't afford a bat and ball. He idolized Jackie Robinson and attended Central High School as a teen. He played semi-pro ball with the Mobile Black Bears, and on November 20, 1951, he signed a contract with the Indianapolis Clowns of the Negro American League. Soon after his standout play as a shortstop with the Clowns, he received offers to play MLB baseball from both the New York Giants and the Milwaukee Braves. He ultimately opted for the Braves.

In his first season, in 1952 with the Eau Claire Bears, the Braves-C farm club, he had an outstanding season. This saw him promoted to the Jacksonville Braves, their Class A farm team, in 1953. As one would expect in those times, being one of the first African Americans to play in the league, he encountered an avalanche of racial prejudice. Encouraged by the manager, Ben Geraghty, and Hank's brother Herbert, Jr., he endured and hung in. That same year, he married his first wife, Barbara Lucas, with whom he had five children. That winter he played in Puerto Rico, where he refined his batting stance under manager Mickey Owen.

Hank Aaron's Grave

Attending spring training with the Braves in 1954, Aaron performed so well that he was signed to a major league contract on the final day of spring training and was assigned uniform number 5. He later switched it to what he called lucky number 44, as Hank had hit 44 home runs in four different seasons. He acquired the nickname "Hammerin' Hank" from his teammates for his batting prowess.

Aaron then embarked on a storied 23-year major league career with the Milwaukee and Atlanta franchises. Not only did he hit for power, but he is one of the few players to achieve a 3,000-hit status, ranking third with 3,771.

His quest to overtake Babe Ruth's home-run record of 714 during the summer of 1973 is the stuff legends are made of. A previously sacred and never thought to be broken record was surpassed by Aaron in a flurry of publicity—some good, some bad (including numerous death threats). He fell short on the last game of the year by one home run. The racial vitriol that ensued in the off season was condemned by Babe Ruth's widow, Claire Hodgson, who stated that her husband would have cheered Aaron's pursuit of the record. It finally came in his home ballpark of Atlanta on April 8, 1974, against the Los Angeles Dodgers. As fans cheered wildly, Hank's parents ran onto the field.

Aaron's final home run, number 755, was hit on July 20, 1976 in

Milwaukee's County Stadium with Aaron in a Brewers uniform. It stood until broken in 2007 by Barry Bonds. Aaron made a surprise appearance on the Jumbotron video screen at AT&T Park in San Francisco to congratulate Bonds on his achievement.

His post-playing career included being a vice president of player personnel for the Atlanta Braves, senior vice president and assistant to the Braves president and corporate vice president for community relations for Turner Broadcasting. In 1982 he was elected to the Baseball Hall of Fame, gaining 97.8 percent of the ballot, second only to Ty Cobb at 98.2 percent.

In 1971, he had divorced his first wife, Barbara, and in 1973 married Billye Suber Williams, with whom he had a daughter. They lived in the Atlanta are, where he died in his sleep in his residence on January 22, 2021, at the age of 86. After his funeral at the Friendship Baptist Church in Atlanta, he was interred at South View Cemetery.

FUN FACTS

The differential that made Hank Aaron sign with the Milwaukee Braves instead of the New York Giants was a mere $50 a month. He had said that the only thing that had kept him and Willie Mays playing together was $50.

Hank Aaron's lucky 44 number was even present when he hit his historic 715th home run. It was off Dodgers pitcher Al Downing, who wore number 44 also.

CEMETERY

South View Cemetery
1990 Jonesboro Road SE
Atlanta, GA 30315
Tel.: 404-622-5393
Hours: Daily, 9 AM–5 PM

DIRECTIONS TO GRAVE

Make an immediate right after entering the cemetery gates. Go approximately 2/3 down the road and Aaron's grave will be on your right.

MABEL FAIRBANKS
Ice Skater
Born: November 14, 1915
Died: September 29, 2001

Mabel Fairbanks was the first African American or Native American to be elected to the U.S. Figure Skating Hall of Fame, in 1997. She was later inducted into the International Women's Sports Hall of Fame in 2001. As a young girl of 10 years, Fairbanks had made up her mind that she wanted to be an ice skater. Fairbanks was initially self-taught by observing skaters at the ice rink in Central Park in New York City. The first pair of ice skates she purchased were at a pawn shop, but they were two sizes too big. She proceeded to stuff them with cotton to make them fit her feet. Through grit and determination, Fairbanks would practice for hours on a tiny ice rink that was constructed by her Uncle Wally. It is sad to reveal that Fairbanks faced discrimination, as she was denied the right to use local ice-skating rinks due to her being black. But she was persistent, and the scrappy youngster was finally allowed by the manager of Gay Blades Ice Rink to use the rink in its final half hour before closing. Instructors at the rink were so

Mabel Fairbanks' Grave

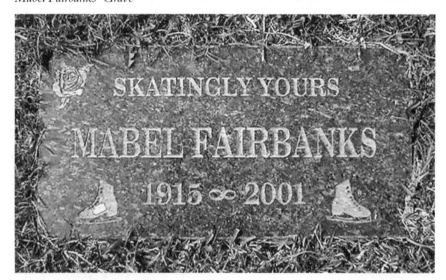

impressed by this girl's dogged determination to learn how to improve her technique on the ice that they would give her tips and instruction free of charge.

Fairbanks toured internationally in the 1940s and 1950s with both the Ice Capades and the Ice Follies. She was often the only black skater in the troupe. She would wow audiences with a move Fairbanks herself created, a spin that involved her holding one leg up, grabbing her foot and stretching the leg behind her back.

Despite her enormous talent, she was denied the opportunity to compete in the Olympics due to the color of her skin. However, Fairbanks would eventually make the Olympics feel her presence and influence. She became trainer and coach to such Olympic winners as Kristi Yamaguchi, Atoy Wilson, Scott Hamilton and Tai Babilonia. Skaters Peggy Fleming and Debi Thomas credit Mabel Fairbanks for influencing and inspiring them.

Mabel Fairbanks passed away in 2001 as the result of leukemia.

FUN FACTS

Fairbanks was inspired to persevere in her pursuit of excellence in her ice-skating after viewing Norwegian skater Sonja Henie in the 1936 movie *One in a Million*.

Mabel had a unique way of signing her autograph. Instead of putting "Best Wishes," Mabel signed "Skatingly Yours."

CEMETERY

Hollywood Forever Cemetery
6000 Santa Monica Blvd.
Los Angeles, CA 90030
Tel.: 323-469-1181
Hours: Dawn to dusk, 7 days a week

DIRECTIONS TO GRAVE

Mabel reposes in the Garden of Legends section. Upon entering the cemetery from Santa Monica Boulevard, drive past the cemetery offices and make a hard left turn before the divider in the road. Take it to the first intersection and you will see a lake with a large mausoleum in the middle of it. Turn right at this road (Nelson Eddy Drive). About halfway up across the lawn, you will see a small bridge leading to the large mausoleum. Mabel's flat marker is at the left of the small bridge before you enter it.

VINCE LOMBARDI
Football Coach/Football Hall of Fame
Born: June 11, 1913
Died: September 3, 1970

In addition to being considered by many the greatest football coach in NFL history, Vince Lombardi is also considered to be one of the greatest leaders and coaches in the history of American sports. That's quite a lot to say, but Vince's accomplishments back those statements up. He taught that whatever you do, give 100 percent, and continually work on improving that 100 percent.

Vincent T. Lombardi was born to first-generation Italians Harry and Matilda, who had settled in Sheepshead Bay, Brooklyn. He initially worked in his father's butcher shop, but felt he had a calling to the priesthood, and enrolled at Cathedral Prep after grammar school. After four years he decided not to heed the calling, however, and enrolled at St. Francis Prep. Upon graduation he received a four-year football scholarship to Fordham University. In his senior year (1936), he was the right guard for the famed "7 Blocks of Granite," which was Fordham's offensive line.

After college, Vince sold insurance and was a debt collector until he got a teaching/coaching position at St. Cecilia High School in Englewood, NJ in 1939. This allowed him to marry his sweetheart, Marie Plainitz. In 1948, he became assistant football coach at his alma mater, Fordham. The next year, he joined West Point Military Academy as an assistant coach under famed coach Earl "Red" Blaik. In 1954, he left to join the New York Giants as assistant coach in charge of offensive coordination for head coach Jim Lee Howell. The Giants fared very well under Lombardi's offensive tutelage, but now in his mid-40s, Vince wanted to be a head coach. There was no movement in the foreseeable future with the Giants. In 1959, the opportunity became available with an offer from the Green Bay Packers, a team mired in last place for many years and in a market ranked last in the NFL. Vince

Vince Lombardi's Grave

jumped at the chance. He inherited a team that had won only one game the previous year.

In his first season, he immediately improved Green Bay to the winning ways that they would get used to under his coaching prowess, posting a 7–5 record. The next year, they made the championship game but lost a heartbreaker to the Philadelphia Eagles 17–13 in the last minutes. Lombardi told the team afterwards, "We will never lose another championship." And they never did.

Lombardi never had a losing record as a coach in the NFL. His stats are impressive. In the regular season he had a 73.8 winning percentage with a 96–34–6 record. Post-season was even more impressive at 90 percent with a 9–1 record. He led the Packers to three straight and five total NFL championships in seven years, as well as winning the first two Super Bowls in 1966 and 1967. John Mara has stated that his father, Wellington Mara, the owner of the Giants, said that the biggest mistake in his life was letting Vince Lombardi leave.

Perhaps his most famous game as a Packer coach was dubbed "the Ice Bowl" because it was played in minus-13–degree weather against the Dallas Cowboys in the NFL championship game at Green Bay on December 31, 1967. They won on the last play of the game, a wedge play that allowed Bart Starr to score the TD.

An innovation that Lombardi came up with for his team was the "Packer Sweep," where two guards paved the way and surrounded the fullback to "Run to Daylight" down the field.

After winning his second Super Bowl, Vince gave up the coaching reins and became general manager of the Packers. However, he got the itch to coach again, and was lured to the Washington Redskins in 1969 by owner Edward Bennett Williams. Once again, Lombardi inherited a last-place team that had not had a winning season since 1955. Under Vince, the Redskins, the perennial doormat of the league, posted their first winning season of 7–5–2.

But tragedy struck Vince in the preseason of 1970, when he was diagnosed with a virulent form of colon cancer. In late June he was admitted to Georgetown University Hospital and succumbed to the disease on September 3. His funeral was at St. Patrick's Cathedral and was officiated by Terence Cardinal Cooke, who delivered the eulogy. Burial followed at Mount Olivet cemetery in Red Bank, New Jersey.

Vince Lombardi was posthumously inducted into the NFL Football Hall of Fame in 1970. The Super Bowl trophy was also christened the Vince Lombardi trophy in his honor.

FUN FACTS

Lombardi was an extremely religious man and he attended Catholic mass daily. He told his team they should live their lives devoted to "religion, family, and the Green Bay Packers," in that order.

Vince Lombardi is perhaps most famous for his saying, "Winning isn't everything, it's the only thing." The fact is that he never uttered that statement. It came from a movie entitled *Trouble Along the Way*, starring John Wayne as a football coach. Wayne's young daughter is the one who voices those words. What Vince did say was, "Winning is not a sometime thing, it is an all-time thing. You don't do things right once in a while . . . you do them right all the time."

CEMETERY

Mount Olivet Cemetery
100 Chapel Hill Road
Red Bank, New Jersey 07701
Tel.: 732-741-5516
Hours: 8 AM–4:30 PM

DIRECTIONS TO GRAVE

Enter the cemetery across from the 100 Chapel Hill Road stay straight for a bit, then bear right and take the next left. Proceed for the length of a football field, and then on the left in section 30 next to the road is the Lombardi grave, next to a crab apple tree.

WILMA RUDOLPH

Olympic Athlete/Gold Medalist
Born: June 23, 1940
Died: November 12, 1994

Wilma Rudolph was a true profile in courage. She overcame childhood polio to become an Olympic gold medal winner in track and field. She was recognized as the fastest woman in the world in the 1960s.

She was born prematurely to Blanche and Ed Rudolph in Clarksville, Tennessee. Her dad was a railroad porter, and her mother worked as a maid.

Wilma contracted polio meningitis at the age of five. She recovered from the polio but lost the strength in her left leg and foot. This necessitated wearing a leg brace for the early part of her life. Her parents sought treatment for her at the Meharry Medical Center in Nashville. For two years, she and her mother made the fifty-mile trip twice weekly. She also received massage treatments four times a day from family members and wore an orthopedic shoe to support her foot for another two years. By the time she was twelve years old, she could walk without the leg brace and shoe support.

Initially home schooled, she attended Cobb Elementary School in Clarksville when she was seven years old. Rudolph then attended Clarksville's all-black Burt High School, where she excelled in basketball and track. In 1958 she enrolled in Tennessee State University. She continued to compete in track and graduated with a BA in education. Her education was financed through a work-study scholarship program that required her to work on campus for two hours per day.

While playing for her high school basketball team, she was spotted by Ed Temple, Tennessee State's track and field coach. When he saw her as a tenth grader, he knew he had seen a natural athlete. Under

Wilma Rudolph's Grave

his guidance, she trained regularly at TSU as a high school student. As a junior, Rudolph attended the U.S. Olympic track and field trials in Seattle, Washington and qualified for the 200-meter individual event, becoming the youngest member of the Olympic team in 1956. Although losing the first event, she ran the third leg of the 4×100-meter relay. The team won a bronze medal. Upon her return to the States, Wilma told friends she had her sights set on the gold at the 1960 Olympic games in Rome.

While a sophomore at TSU, she competed in the U.S. Olympic track and field trials at Abilene Christian University in Abilene, Texas. There, she qualified for the 1960 Summer Olympics in the 100-yard dash.

At the 1960 Olympic games in Rome, Wilma realized her dream of winning Olympic gold in resounding fashion. In the 100-meter and 200-meter sprints and 4×100-meter relay, she won a gold medal in each of these events, and became the first woman to win three gold medals. She ran the anchor leg of the relay race and overcame a lead by Germany in a very close finish. Rudolph was hailed, as said, as "the fastest woman in the world" and as a result became an international star.

Back home, she appeared on the *Ed Sullivan Show* and the game show *To Tell the Truth*, which bolstered her fame. In 1961, she competed in the indoor track circuit in the Milrose Games, the Penn Relays, and New York Athletic Club track events. She decided to retire from active competition and not compete in the 1964 Olympic games in

Tokyo, preferring to exit at the top of her game. She felt that not duplicating her trifecta of medals would be a letdown.

In May 1963, after returning from a good-will tour in West Africa on behalf of the U.S. State Department, she participated in a civil rights protest in her hometown of Clarksville to desegregate city restaurants. In a short time, it proved successful.

Wilma was married twice, first in 1961 to William Ward, a North Carolina college track team member. They divorced two years later. After graduation from TSU in 1963, she married her college sweetheart, Robert Eldridge, with whom she already had a daughter. The marriage produced four children, but they divorced 17 years later.

After retirement, Wilma turned to teaching, starting as a second grade schoolteacher at the Cobb Elementary School of her youth. She subsequentially had posts in Indianapolis at the Federal Job Program and at DePauw University as the director of the women's track team. She also had a local TV show in Indianapolis. In 1961, she established the Wilma Rudolph Foundation to train young athletes. She was a TV sports commentator at the 1984 Olympic games in Los Angeles for ABC. Two years prior to her death in 1994, she became a vice president at Nashville's Baptist Hospital.

In July 1994 she was diagnosed with both throat and brain cancer. Her condition worsened quickly, culminating in her death in early November. She expired at her home in Brentwood, a suburb of Nashville, at the young age of 54. She was interred in the Edgefield Missionary Baptist Church Cemetery in her hometown of Clarksville, next to her mother.

FUN FACTS

As a result of her fame at the 1960 Olympics, she received some colorful nicknames. The Italians dubbed her "La Gazella Nera" (the black gazelle). The French called her "La Perle Noire" (the black pearl).

In the early 1960s Wilma dated her fellow Olympian Cassius Clay, later known as Muhammad Ali.

CEMETERY

Edgefield Missionary Baptist Church Cemetery
503 Tompkins Lane

Clarksville, TN 33704
Tel.: 931-647-2776
Hours: Mon.–Sun., 8:30 AM–4 PM

DIRECTIONS TO GRAVE

Enter the cemetery between the brick pillars of 1400 Paradise Hill Road, and Wilma's grave is in the center loop of the drive.

CHAPTER SIX

Entertainment

HARRY HOUDINI

Magician/Escape Artist
Born: March 24, 1874
Died: October 31, 1926

Harry Houdini was America's first entertainment superstar. In the early twentieth century, he entranced the nation by his feats of slight-of-hand, his uncanny escapes, and his aviation exploits. He was center stage in the nation's craving for excitement as the new century took hold. However, it was his crusade to debunk spiritualism that led us to include Houdini in this volume of American heroes.

Born Erik Weisz in Budapest, Hungary, Harry's parents were Rabbi Samuel Mayer and his wife Cecelia. His family immigrated to the United States in 1878, and they initially settled in Appleton, Wisconsin. They moved to Milwaukee in 1882. Finally, in 1887 in search of a better future, the family relocated to New York City. Growing up, Erik was a prolific runner who won many medals.

Harry Houdini's Grave

As a teenager, Erik also took to magic and adopted the stage name of Harry Houdini after his idol, the French magician Robert Houdin. Harry probably came from "Ehri," his nickname. He began his magic career in 1891 in tent shows, initially with card tricks, billing himself "King of Cards." He performed with his brother "Dash" (Theodore) at the Chicago World's Fair in 1893 before returning to New York City to perform at Huber's Dime Museum. There, he met a fellow performer, Wilhelmina Beatrice "Bess" Rahner, whom he married. They performed together as "the Houdinis."

Houdini's big break came in 1899 when he met vaudeville empresario Martin Beck, who was impressed by his handcuff escape act. He then booked Harry on the Orpheum vaudeville circuit, and this was followed by Houdini touring Europe the next year. The tour was a smashing success, with Harry escaping from all types of constraints and jails in Great Britain, the Netherlands, France, Russia and Germany. He was now earning $300 per week. In 1904, he returned to the United States and bought a townhouse at 278 West 113th Street, Harlem, which became his home for the rest of his life.

From 1900 through the 1910s, he performed with great success in Great Britain and the United States. Many towns saw him suspended upside down and lofted up high by a crane over the street as he escaped from a strait jacket. He invited the public to think up contraptions to hold him including being chained, nailed into a packing crate and lowered into the water. Perhaps his most famous escape was the Chinese Water Torture Cell, a version of the Milk Can Escape. In this trick, he was locked upside down in a glass and steel cabinet overflowing with water as he held his breath. He performed this trick for the rest of his life. A famous non-escape trick was making an elephant disappear on stage, which he did on the stage of the Hippodrome Theater in New York City in January 1918.

Houdini became enamored of flying, then a nascent activity, and he purchased a biplane in 1909 for $5,000. After crashing once, he successfully flew on November 26 in Hamburg, Germany. The following year while touring Australia, he made three successful airborne flights, the last one lasting seven and half minutes and covering six miles.

In 1918 Houdini took a stab at the film industry, making three. The best known one was *The Grim Game*. Alas, his acting career did

Signature of Houdini, from the private collection of Vincent Gardino

not take off, and in 1923 he gave up on the movie business, stating that "the profits are too meager."

As spiritualism had taken root in America in the 1920s, Houdini dove in and debunked reported mediums who preyed on bereaved relatives who trusted them to make dead loved ones reappear or communicate with the living. His most renowned unmasking was the medium Margery Crandon in Boston. A famous proponent of spiritualism was the author Sir Arthur Conan Doyle, creator of super-sleuth Sherlock Holmes. Houdini crossed swords with him on a few occasions, showing him how all the supposed appearances and rappings were just illusions and tricks.

In early October, 1926 in his dressing room at the Princess Theater in Montreal, as Houdini reclined on a sofa due to a broken ankle, he was struck in the abdomen by a student, Jocelyn Gordon Whitehead, several times without having time to brace himself for the hammer-like blows below the belt. Throughout the evening he was in great pain, but performed anyway. He finally saw a doctor, who said he had a temperature of 102 degrees and a diagnosis of acute appendicitis. Told to have immediate surgery, he nevertheless pressed on to perform at the Garrick Theater in Detroit on October 24. He passed out at the show, but was revived to complete his final performance. Then hospitalized, his subsequent operation uncovered the fact that deadly poisons from the appendicitis had infected his entire body. He died appropriately on Halloween. His funeral was held on November 4 in New York with over 2,000 in attendance. Interment followed in Machpelah Cemetery in Queens in the impressive Houdini plot with his bust placed at the entrance of the cemetery.

After his death, numerous seances were conducted by his widow Bess and the medium Arthur Ford. Harry had given Bess a code, "Rosabelle believe." After ten years of holding the seances on Halloween, Bess ended them, stating "ten years is enough to wait for any man." His widow and assistant died of a heart attack in Needles, California on a train enroute to New York. Despite the fact that she had expressed her desire to be buried beside Harry, she was interred in Gate of Heaven Cemetery in Westchester, as her Catholic family refused to have her buried in a Jewish cemetery.

FUN FACTS

Harry Houdini himself was a "grave tripper." He made it a point to visit fellow magicians' graves and in many cases rehabilitated their sites that had fallen into disrepair.

Harry served as the president of the Society of American Magicians from 1917 to 1926. In addition, he was also president of Martinka & Company, America's oldest magic company. It is still in operation today in Midland Park, NJ.

CEMETERY

Machpelah Cemetery
8230 Cypress Hills Street
Ridgewood, NY 11385
Hours: 10 AM–dusk

DIRECTIONS TO GRAVE

Upon entering the small cemetery, you cannot miss the impressive Houdini family plot with the bust of Harry in the center. Harry's grave is to the left in the plot.

JUDY HOLLIDAY

Actress/Academy Award Winner
Born: June 21, 1921
Died: June 7, 1965

Judy Holliday was an actress, singer, and comedienne who displayed exemplary courage in her prolonged battle with cancer at a relatively young age. As a performer, she was honored with an Academy Award, a Golden Globe and a Tony.

She was born to Abe and Helen Tuvim in New York City. Her parents were of Russian Jewish descent. Her father was a political

Judy Holliday's Grave

activist and her mother taught piano. She grew up in Queens, New York and graduated from Julia Richman High School in Manhattan. Her first job was as a telephone operator for the Mercury Theater.

Her show business career began as part of a nightclub act of a troupe called "the Revuers." They played their engagements at the Village Vanguard, the Blue Angel and the Rainbow Room. In 1940 they released an album entitled *Nightlife in New York*. In 1944 the Revuers filmed a scene for a Carmen Miranda movie entitled *Greenwich Village*. That same year, at the age of 22, Judy scored her first film role as an airman's wife in 20th Century Fox's *Winged Victory*. This was despite the advances of studio head Darryl Zanuck, which she rebuffed. The following year she made her Broadway debut on March 20 at the Belasco Theater in *Kiss Them for Me*. She received the Clarence Derwont Award for "most promising female actress." In 1948, she married David Oppenheim, who became a classical and television producer. The union produced a son, Jonathan.

In 1946, she returned to Broadway to play the role that launched her fame as the scatter-brained Billie Dawn in *Born Yesterday*. Columbia bought the screen rights to adapt the film, and based on her rave Broadway reviews and key appearance in the Hepburn/Tracy film *Adam's Rib*, studio boss Harry Cohn signed her for the part. In 1950, she won the Golden Globe and edged out experienced actresses Gloria Swanson, Eleanor Parker, Bette Davis, and Anne Baxter for the Academy Award for best actress.

That same year she had to endure an investigation by the Pat

McCarren Senate Internal Security Subcommittee. This was because her name appeared as one of 151 "pro-communist" activists in the conservative publication *Red Channels*. She bravely testified that she abhorred Stalinism and authoritarianism but defended free speech. She escaped relatively untarnished from her testimony.

In 1954, she starred in a number of motion pictures, two of them being newcomer Jack Lemmon's films *It Should Happen to You* and *Phfft*. *The Solid Gold Cadillac* was released in August of that year starring Judy opposite Paul Douglas. Later that year, she returned to Broadway in *Bells Are Ringing*, directed by Jerome Robbins, for which she won the Tony Award for best actress in a musical. The *New York Times* theater critic Brooks Atkinson wrote, "She has gusto enough to triumph in every music hall antic."

She returned to the screen in 1960 in her last movie, opposite Dean Martin, reprising her role in in the Broadway hit *Bells Are Ringing*. That year, she was awarded a star on the Hollywood Walk of Fame.

Judy began her heroic battle against the then dreaded disease of cancer in 1960 when it was discovered that she had throat cancer. In the same operation to deal with her throat, it was discovered that she also had breast cancer, which necessitated a mastectomy. Her battle was waged against an environment in the 1960s that cast an aura of shame and revulsion on the disease. Despite this, Judy underwent her painful chemo and radiation treatments while still raising her young son Jonathan. She was also very open about her disease despite the taboos surrounding it. She filmed commercials for the American Cancer Society requesting contributions for research.

Her last acting role was in the stage musical *Hot Spot*, which ran for five months in 1963. Toward the end of its run, it was found that the cancer had returned to her throat. She retreated to her home in the country in Washingtonville, New York. Five years after her diagnosis and after countless treatments (toward the end she was receiving heroin to dull the pain), she still continued to care for her son until a month before she died. Her final work was composing the music for the film *A Thousand Clowns* in 1965. After a valiant fight, she finally succumbed to metastatic breast cancer at Mt. Sinai Hospital. Her funeral was held at Frank Campbell Funeral Home with interment following at Westchester Hills Cemetery in Hastings-on-Hudson, New York.

FUN FACTS

Judy derived her name Holliday from the translation of her Jewish name, Tuvim.

Despite Judy's reputation in the cinema as being scatter-brained, she boasted an IQ of 172.

CEMETERY

Westchester Hills Cemetery
400 Saw Mill River Road
Hastings-on-Hudson, NY 10706
Tel.: 914-478-1767
Hours: 8:00 AM–4 PM; closed on Saturday

DIRECTIONS TO GRAVE

When you enter the cemetery, drive up a short bit and you will see the tomb of Broadway empresario Billy Rose. Make a right and you will see the Gershwin tomb, which is the third one on the right. The last tomb on that row is Guggenheim. Continue straight ahead past that mausoleum and another one on the left. Approximately 100 feet straight ahead on the left, you will see the stone marker Tuvim. In front of that is Judy's foot stone.

MARY TYLER MOORE
Actress
Born: December 29, 1936
Died: January 25, 2017

Some of you may be wondering how we can include Mary Tyler Moore as a heroine in this volume of *Grave Trippers*. We believe Mary Tyler Moore was an actress and producer who changed the way women viewed themselves in the 1960s and '70s. She swept away the stereotype of a woman as primarily a homemaker and introduced, via her portrayal in *The Mary Tyler Moore Show*, the fact that a woman can be successful both being single and having a career. This helped define a new vision of American womanhood.

Mary was born in Brooklyn Heights, New York to George and

Mary Tyler Moore's Grave

Marjorie Moore. Growing up in an Irish Catholic household, she attended parochial schools. When she was eight years old, the family moved to Los Angeles. Early on, Mary knew she wanted to be a performer. Her TV career began as a tiny elf in Hotpoint appliance commercials during the 1950s. She auditioned for the part of Danny Thomas's daughter in his television show and narrowly missed getting the part.

Her first regular television role was as a mysterious and glamorous phone operator in the series *Richard Diamond, Private Detective* starting in 1957. She was never seen except in shadow, so the audience did not know what she looked like. After three seasons, she went in and asked for a raise and was immediately fired. She then guest-starred in many television series during that time.

Her first big break came in 1961 when Carl Reiner cast her as the wife of Dick Van Dyke in *The Dick Van Dyke Show*. As Laura Petrie, she won rave reviews for her comedic performances and became internationally famous. She began to break the stereotypical view of women by being the first actress to wear pants instead of a dress. Her capri pants became her trademark look. The show ran till 1966.

In her personal life, Mary had married Richard Meeker in 1955 and they had a son, Richard Meeker, Jr. They divorced in 1962. Later that year, Mary married Grant Tinker, a CBS executive.

In 1970, Tinker and Moore pitched a TV sitcom to CBS about a single woman in a TV newsroom. This bridged aspects of the women's movement and mainstream culture by portraying an amiable single woman focused on her career rather than marriage and family. The half-hour show *The Mary Tyler Moore Show* was an instant hit and became the anchor of CBS programming on Saturday night. It ran for seven seasons and solidified Moore's reputation as a true star. It was also the centerpiece of Moore and Tinker's production company, MTM Enterprises. Mary was responsible for the creative end while Tinker handled all business aspects. Their collaboration continued until their divorce in 1981.

After the series' end, Mary continued to work in other TV series as well as theater and film. Her most famous film, for which she received an Oscar nomination, was the coming-of-age drama *Ordinary People*.

Mary experienced a great tragedy in her life when her son accidently shot himself in 1980 with a shotgun that had a hair trigger. This spiraled her drinking and led to her checking herself into the Betty Ford Clinic. In addition, her diabetes, which was initially diagnosed in 1969, flared up. Mary became a tireless advocate of diabetes awareness using her celebrity status to bring this issue to the forefront of Americans.

In 1982, Mary met the love of her life, cardiologist Dr. Robert Levine, and they were married the next year despite an 18-year age difference. They remained devoted to each other for 34 years, ending with Moore's death in 2017.

Mary's health issues became serious in the 21st century. In 2011 she had surgery to remove a benign brain tumor. She also had heart and kidney problems and the diabetes that ravaged her body and left her nearly blind in her final years.

Surrounded by her loving husband and close friends, Mary died on January 25, 2017, in Greenwich Hospital in Greenwich, Connecticut, the result of pulmonary arrest complicated by pneumonia. She was interred in a private ceremony in Oak Lawn Cemetery in Fairfield, Connecticut.

FUN FACTS

The reason Mary missed out on the role of Danny Thomas's eldest daughter was that her nose was too small. Thomas said that no daughter of his could have had a nose that small.

The logo of Mary's production company, MTM Enterprises, resembled the MGM logo but had a kitten instead of a lion.

Many of the major cast members went on to greater fame after the series *The Mary Tyler Moore Show* concluded, such as the following:

> Cloris Leachman for the spinoff *Phyllis*
> Valerie Harper for the spinoff *Rhoda*
> Ted Knight for *Too Close for Comfort*
> Ed Asner for *Lou Grant*
> Gavin Macleod for *The Love Boat*
> John Amos for *Good Times*
> Betty White for *The Golden Girls*
> Georgia Engel for *Everybody Loves Raymond*

CEMETERY

Oak Lawn Cemetery
1530 Bronson Road
Fairfield, CT 06824
Tel.: 203-259-0458
Hours: 9 AM–5 PM, except summer till 7 PM

DIRECTIONS TO GRAVE

At cemetery entrance, follow Cemetery Road to Maple Road, where you will make a left. Maple Road will take you to Area "D," which is the third section on the left. When you see the Peck monument on the left, take that left and walk down the path. On the left will be an angel, but that is not Mary's grave. About 10 to 15 feet further on the right, you'll see another one, and you will have found Mary's resting place overlooking a lake. Her site faces a cluster of trees.

ANNA MAY WONG
Actress
Born: January 3, 1905
Died: February 3, 1961

Born Wong Liu Tsong, Anna May Wong is known as the first Chinese American actress and movie star in Hollywood to also attain international acclaim. Her versatility is evident in her career, encompassing silent and sound films, television, radio, and stage.

Anna's parents were second-generation Taiwanese Chinese Americans who settled on Flower Street in Los Angeles and later moved to Figueroa Street. There, they were the only Chinese people on the block. She attended public school with her older sister and was subject to racial taunts. Classes were in English, but Anna attended a Chinese language school on afternoons and weekends.

Anna became infatuated with films at a very young age. In 1919, a friend got her a role as an extra in the Metro production of *Red Lantern* and thus set her on the road to her cinematic career. Finding it difficult to go to school and pursue a movie career, she dropped out of Los Angeles High School in 1921. Her first screen credit was that year in *Bits of Life* opposite Lon Chaney. The *New York Times* gave her a favorable review, saying, "she should be seen again and often on screen."

Despite this, Hollywood proved reluctant to create starring roles for Wong, relegating her to supporting roles such as an exotic concubine in 1923's Tod Browning production of *Drifting*. At the age of 19 the following year, she was cast as a stereotypical "dragon lady" (strong, deceitful, domineering, mysterious and sexually alluring) in *The Thief of Bagdad* opposite Douglas Fairbanks. Her brief appearance helped to introduce her to the general public. She began a relationship with Tod Browning, who had directed her in *Drifting*. But Wong continued to be offered "vamp" stereotype roles in the cinema.

Not wanting to being typecast, Wong left for Europe in 1928. There she became a sensation. In Vienna she played the title role in the operetta *Tschun Tachi* in fluent German. A German critic remarked, "Her acting was deeply moving, carrying off the difficult German speaking part very successfully." In 1929, she made her last silent film, *Piccadilly*, where she had a starring role that accentuated her sensual-

Anna May Wong's Grave
(Courtesy of Susan Lukenbill)

ity. Wong received many plaudits from the critics in England. *Variety* commented that she "outshines the star" (Gilda Gray) and *Time* called it her best film.

In 1930, she returned to the U.S. and was promised a movie contract with leading roles and top billing by Paramount. Wong appeared opposite Marlene Dietrich as a self-sacrificing courtesan in 1932's *Shanghai Express*. Her sexually charged scenes with Dietrich sparked rumors of a relationship. Today many film historians feel that Wong's performance was superior to Dietrich's. Once again disappointed with Hollywood's lack of cinematic support, she traveled back to Great Britain in the mid '30s, where she starred in films and vaudeville shows.

The popularity of Pearl S. Buck's book *The Good Earth*, which highlighted China's struggles with the Japanese, caused MGM to cast a movie based on the book. The movie was made in 1936, but was released in 1937. It unveiled an incredible snub against Anna, who had openly desired the lead role of Olan. Instead, Irving Thalberg of MGM cast Louise Rainer in the role of Olan. Thalberg had offered a lesser role to Anna as Lotus, a deceitful song girl, which she refused. Louise Rainer went on to the win the Oscar for best actress for that role—a notorious example of casting discrimination in the 1930s. The most likely reason why Thalberg casted a white actress instead of Wong was due to the Hays Code anti-miscegenation guidelines forbidding any implied sexual relations between different races. Paul Muni, a white actor, was cast in the Asian main lead role.

After this major disappointment, she embarked on a yearlong tour of China in 1936 to visit her father and family. After returning to Hollywood in the late 1930s, Paramount cast her in "B" movies, and she used them to her advantage to portray successful professional Chinese American characters. She also tutored other actors such as Dorothy Lamour as a Eurasian in *Disputed Passage*. Wong's career continued into the 1940s just as before, portraying stereotypes.

In 1951, Wong starred in the TV detective series for the Dumont Television Network called *The Gallery of Madame Liu Tsong*, using her birth name. Ten half-hour shows aired, but it was not renewed for a second season. In the 1950s, she did guest appearances on television shows such as *Adventures in Paradise*, *The Barbara Stanwyck Show* and *The Life and Legend of Wyatt Earp*. She returned to film in *Portrait in Black* starring Lana Turner, and yet again found herself stereotyped.

Anna suffered from recurring health problems through the 1950s. She was scheduled to play Madame Liang in Oscar and Hammerstein's *Flower Drum Song*, but couldn't accept the role due to her health issues. On February 3, 1961, she died of a heart attack in her sleep at her home in Santa Monica, two days after her final performance in *The Barbara Stanwyck Show* in an episode entitled "Dragon by the Tail." She was cremated and interred beside her sister and mother in Angelus Rosedale Cemetery.

She attained a star in the Hollywood Walk of Fame in 1960, the first Asian American actress to be so honored. She is also one of the supporting pillars of the "Gateway to Hollywood" sculpture alongside those of Dolores Del Rio, Dorothy Dandridge and Mae West on Hollywood Boulevard. In 2023, the U.S. Mint produced over 300 million quarters with her likeness on the back as a part of the American Women Quarter series.

FUN FACTS

In 1926, Wong put the first rivet into the structure of Grauman's Chinese Theater when she joined actress Norma Talmadge for the groundbreaking ceremony.

In addition to her acting prowess, in the 1930s she was seen as a fashion icon for over a decade. In 1934, the Mayfair Mannequin Society of New York voted her the "World's Best Dressed Woman." In 1938, *Look Magazine* named her "the world's most beautiful Chinese girl."

CEMETERY

Angelus Rosedale Cemetery
1831 W. Washington Blvd.
Los Angeles, CA 90007
Open: Daily 8 AM–4:30 PM

DIRECTIONS TO GRAVE

Enter cemetery at Washington Blvd. entrance and turn right immediately at the first road. Continue on the path east, then north and take the right fork at the intersection. Look for the small black "N5" marker at the edge of the road, and the "Stocker" mausoleum. Anna May is interred six rows away and approximately 100 feet from the road under a pink headstone bearing her mother's name, Lee Toy Wong. Anna's name, as well as her sister's, is in Chinese.

HATTIE McDANIEL

Actress/Academy Award Winner
Born: June 10, 1893
Died: October 26, 1952

Hattie was a trailblazing woman of color who became the first black actress to win an Academy Award in 1939. The feat would not be repeated until 50 years later. Along the way, she developed a legendary reputation for kindness and generosity (not seen often in Hollywood) by lending money to friends and strangers alike.

She was born in Wichita, Kansas, the youngest of 13 children, to former slaves Henry and Susan McDaniel. In 1898 the family moved to Denver, Colorado.

At an early age, Hattie demonstrated a capacity for singing and dancing. Her brother Otis wrote a well-received play, *Champion of the Freedman*, that ran in Kansas City, Missouri. Hattie joined him when she was 15 years old and quit high school. She joined the McDaniel's

Hattie McDaniel's Grave (Courtesy of Susan Lukenbill)

Stock Company and appeared as a singer-comedienne. She also played Mammy in a head rag and sang black-face at church gatherings, tent shows and socials. She couldn't sustain a living as a performer, however, and was forced to work as a domestic.

In 1925, Hattie found herself back in Denver still working as a maid. She tried Chicago soon after, but found that work there was dwindling due to audiences moving away from vaudeville and to motion pictures, which continued to grow in popularity. Meanwhile, Hattie's brother Sam had found work in Los Angeles as a radio actor. Hattie, finding herself parentless and twice divorced, followed suit in 1931.

Once settled in Los Angeles, she ingratiated herself with Charles Butler, a black casting director at Central Casting Corp. He began to cast her as a servant with a salary of $7.50 per day. She soon joined KNX Radio with her own program, *Hi Hat Hattie and Her Boys*, on which she sang and told jokes. In 1934, she got her big break when Director John Ford cast her to play a maid opposite Will Rogers in *Judge Priest*. The well-known gossip columnist Louella Parsons gave her a good review.

Soon she found that with her income in film she could totally support herself. With her newfound money, Hattie bought an elegant house in the West Adams section of Los Angeles. At Christmas, she filled it with gifts for poor black children.

This good will manifested itself with landing the role of Mammy in David O'Selznick's production of *Gone with the Wind* in 1939. The first director, George Cukor, deemed that Hattie lacked dignity for the role but Selznick prevailed, and she got the part, which secured her place in Hollywood history. The ugly era of Jim Crow laws prevented McDaniel from attending the film's premiere in Atlanta, but she was able to do so at the Los Angeles premiere and the 12th annual Oscar ceremonies at the Ambassador Hotel. At the awards, she sat at a segregated table at the side of the room. Louella Parsons praised her acceptance speech as one of the finest ever given on the Academy floor. Humbly, she accepted it for herself and "her race."

Despite now being a celebrated actress with a new contract with David Selznick, she was offered few meaningful roles. She survived by donning her Mammy costume and making personal appearances at white Los Angeles theaters. All the while, she continued being kind to fellow performers and her community. Singer Lena Horne credits her with helping Horne find her way in LA with her kindness. During World War II, she was very active entertaining troops by making personal appearances, throwing parties and performing at the USO.

In 1951, after filming only four episodes of a television show entitled *Beulah*, Hattie learned she had breast cancer and was forced to leave the show. She found herself nearly broke due to her constantly helping other people out. Uncomplaining, she had to sell her grand home and move into Motion Picture Country Home and Hospital, the first African American to do so. She expired there on October 26, 1952. Her estate was valued at less than $10,000 at the time of her death.

Her desire was to be buried at Hollywood Memorial Park (now Hollywood Forever), but it was thwarted by the owner, Jules Roth, who forbade blacks in the cemetery. She was instead interred in her second choice, Angelus Rosedale Cemetery, despite protests even at that cemetery, which had an all-white policy that her family ignored. The new owners of Hollywood Forever in 1999 offered to re-inter Hattie there, but her family demurred. Instead, Hollywood Forever erected a beautiful cenotaph in her memory in a prominent place in the cemetery.

FUN FACTS

Hattie has not one but two stars on the Hollywood Walk of Fame. One is for her radio work, the other for her screen work.

Hattie was well known for her clever punch lines. The best known is, "I would rather make $700 a week playing a maid than being one." Another line was in an interview where she stated that "I've played everything but a harp."

CEMETERY

Angelus Rosedale Cemetery
1831 W. Washington Blvd.
Los Angeles, CA 90007
Tel.: 323-734-3155
Hours: Daily, 8 AM–4:30 PM

DIRECTIONS TO GRAVE

Upon entering the cemetery at the West Washington Blvd. entrance, take your first left after the main gate. Almost on the corner on the right, you will find Hattie buried in the third row about 20 feet in. Her section is identified by a black marker on the corner saying "Palm View Section D."

FRANK SINATRA

Entertainer
Born: December 12, 1915
Died: May 14, 1998

As we indicated in our Introduction, some of our choices for the theme of heroism for inclusion in this book may be controversial. Some of you may be thinking, how is it that a man with a reputation as a hard-drinking womanizer with a hair-trigger temper, as well as with rumors of mob ties, can be considered a hero? The answer lies in Frank Sinatra's generosity.

Some examples of Sinatra's generosity are covered here before we proceed with his biography. Older movie fans will recognize the name of actor Lee J. Cobb. His dominating persona and booming voice often led to his playing a villain in films. He was the corrupt union leader in *On the Waterfront* who clashed with Marlon Brando's character at the end of the film. In the 1950s, Cobb refused to testify before the House Un-American Activities Committee when he was accused of being a communist. This in turn led to a period of forced unemployment due to Cobb being blacklisted. As a result, Lee J. Cobb's wife suffered a nervous breakdown and Cobb found himself in essence bankrupt. When Sinatra heard of his financial straits, Sinatra paid Cobb's wife's hospital bills. This, despite the fact he had never met Lee J. Cobb before. Another actor Sinatra befriended was Bela Lugosi, who played the title role in 1931's classic movie *Dracula*. After reading about the sad story of Bela's recovery from morphine addiction and seeing the pictures of a frail Lugosi, Frank paid Lugosi's hospital bills. And as with Lee J. Cobb, Frank Sinatra had never met Lugosi before.

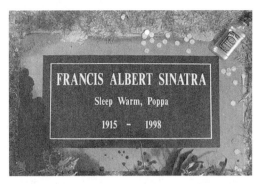

Frank Sinatra's Grave

Sinatra was also a huge supporter of the State of Israel ever since it was founded in 1948. Frank helped the Israeli government raise nearly seven million dollars via pledged bonds. Sinatra was also a champion for equal rights for African Americans as he led the way to boycotting hotels and casinos that were not allowing black patrons and performers entry into their establishments. It is for these reasons that we consider "the Chairman of the Board" a hero.

Francis Albert Sinatra was born in Hoboken, New Jersey to Italian immigrant parents. His was a difficult birth, as the delivering doctor used forceps, which caused permanent scarring and punctured Frank's left eardrum. They feared the baby was stillborn as he was not breathing, but Sinatra's grandmother, who was present, took her new grandson and ran cold water over the baby until he started to breathe. At a young age, he showed great interest in music, and like many of his generation was influenced by the singing of Bing Crosby. He quit high school and began to sing professionally. He teamed up with three other singers and formed "the Hoboken Four." When the group performed on live radio, Sinatra's name began to circulate and his singing, especially among teenage girls (bobbysoxers, as they were referred to), became popular.

In the late '30s and early '40s, he gained valuable experience singing under contract with the bands of Harry James and Tommy Dorsey. As his popularity grew, Sinatra went solo and signed with Columbia Records. Singles such as "People Will Say We're in Love" and "You'll Never Know" were big hits. The demand for Sinatra's records was beginning to rival his fellow crooners such as Perry Como and even Bing Crosby himself.

Frank Sinatra's career path was certainly, up to the mid '40s, approaching a zenith. He was extremely popular and in demand for live performances on the radio and for concerts. However, arguably due to overexposure, less than stellar single releases, stories of ties to mobsters, a messy divorce from his first wife Nancy, and a torrid affair with screen goddess Ava Gardner (whom Sinatra later married), by 1950 Sinatra's career had hit bottom. A Columbia music exec remarked that they couldn't even give away any of Sinatra's records. Few would argue that 1948's *The Kissing Bandit* was probably the nadir of his film career as well.

However, Sinatra's popularity came roaring back when he won an Academy Award as best supporting actor for 1953's *From Here to Eternity*, co-starring Burt Lancaster. Just two years later, Frank put in another strong film performance for his role as a recovering drug addict who falls off the wagon in *The Man with the Golden Arm*. He again received an Oscar nomination for that performance, but lost to Ernie Borgnine, who won for *Marty*. The '50s saw Frank Sinatra reinvigorated musically as well. He signed with Capital Records in 1953. His collaborations with conductors and arrangers Nelson Riddle, Gordon Jenkins and Count Basie, music legends in their own right, produced arguably for Sinatra his strongest tracks with hits such "Young at Heart," "Come Fly with Me," and "Nice 'n Easy." Sinatra's best saloon songs, as he would call them, were also recorded in this time frame such as "Make It One for My Baby" and "Angel Eyes."

Frank was also again in demand for live performances, becoming one of the biggest draws in Las Vegas history. His comradery with fellow "rat packers" Dean Martin and Sammy Davis, Jr. lifted their careers as well, as they often performed together on television and in movies. Though they never came out and said it, of the three, it was implicit that Frank Sinatra was the leader of the pack. Only when Elvis Presley returned to live performances in 1969 was there another "king" in Las Vegas to rival Sinatra.

In 1960, Frank left Capital Records and formed his own label, Reprise Records. During the '60s Sinatra recorded songs that are now considered classic such as "My Way," "Strangers in the Night" and the haunting "It Was a Very Good Year." In addition, Sinatra delivered superb film performances in 1962's *The Manchurian Candidate* and in 1965's *Von Ryan's Express*, which were box office hits.

Frank's marriage to Ava Gardner had long ended when they divorced in 1957. In 1966, he married Mia Farrow. However, that marriage lasted only two years, dissolving in divorce.

At the age of 55, in 1971 Frank Sinatra announced his retirement from show business; however, the retirement was short-lived. He came back to accolades for his new album, *Ol' Blues Eyes Is Back*, in 1973. A television special of the same name in 1973 co-starring his good friend, dancing legend Gene Kelly, was also well received. A highlight of Frank Sinatra's career came in 1974 when he sang at New York City's famed Madison Square Garden. The capacity crowd roared with loving approval from the moment of Frank's entrance to his final song, which was a live rendition of "My Way."

By the end of the 1970s, Sinatra was past his prime, but his voice had aged very well. In 1980, his album *Trilogy* was released. Among the songs included in this three-album set was to be his last big single release, which was the classic "New York New York." Up until the very end when he gave his last concert in 1994, Sinatra sold out every one of his performances.

On May 14th, 1998, Frank Sinatra died at age 82, coincidentally during the airing of the Seinfeld finale, due to a heart attack.

FUN FACTS

Did you know that country music star Glen Campbell played guitar on Sinatra's studio recording of "Strangers in the Night"?

Ironically, Ernest Borgnine, who beat out Sinatra for the Oscar at the 1956 Academy Awards ceremony, also starred with Frank in *From Here to Eternity*.

CEMETERY

Desert Memorial Park
31705 Da Vall Drive
Cathedral City, California 92234
Tel: 760-328-3316
Hours: Mon.–Fri., 7 AM–4:30 PM; closed on weekends and holidays

DIRECTIONS TO GRAVE

Enter the cemetery via Da Vall Drive and make the very first left. You will see near the cemetery office on the left parking. Then you will walk west to Sinatra's grave. His is relatively near the road and often adorned with flowers. On his grave marker is inscribed "Sleep Warm, Poppa." The marker is blackish with white text.

ED SULLIVAN
Variety Television Host
Born: September 28, 1901
Died: October 13, 1974

If you are old enough, you could not be unaware of who Ed Sullivan was. A fixture on Sunday evenings, his variety show aired on television for an impressive 23 years. This show featured the acts of comedians, serious actors, singers from rock n' rollers to grand opera, jugglers, acrobats, dancers and animal acts as well as puppet acts. In our first book, *Grave Trippers: History at Our Feet*, we featured Ed Sullivan in our introduction. We provided a photo of Sullivan's autograph, which was obtained for us by our father, who worked as a waiter at a celebrity-haunt restaurant. That autograph, as well as others our father would bring home, got us interested in researching about those luminaries and this led, in turn, to our love of history.

We consider Ed Sullivan a hero because he was generous to many entertainers, and he was willing to stand up for minority performers, as you will see.

Edward Vincent Sullivan was born in Harlem in New York City, but he was raised in Port Chester, just north of New York City. In high school he was a very talented athlete, playing football, basketball, baseball and track. After graduating, over the next ten years or so he worked at a number of newspapers, mostly as a sportswriter. Sullivan got his big break in 1929 when he joined the *New York Daily News* as a show business gossip columnist. Ed wrote a column titled *Little Old New York*, where he concentrated on writing about Broadway shows and show business news. The column became very popular. Sullivan's popularity led him to hosting a variety radio program in 1941 on CBS.

Ed Sullivan's Crypt

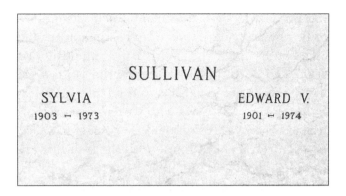

In 1948, as the result of Sullivan's popularity both in print and on the radio, CBS offered Sullivan the opportunity to host a variety television program called *Toast of the Town*. Early TV critics gave the program poor reviews, and most of them were directed not at the acts that appeared on the show but on the host himself. The critics said Sullivan's movements were stiff and awkward and that he had no personality. Yes, the critics were down on Sullivan, but the viewing public loved him. The fact that the show changed its name in the 1950s to *The Ed Sullivan Show* and lasted so long speaks for itself. The viewing public saw Sullivan's mannerisms and distinctive manner of speaking as endearing. In fact, his voice and stiff body movements were fodder for impressionists who would appear on the show. Sullivan usually would laugh along with the audience.

Sullivan often offered the viewing public a first glimpse of individuals who would later become established names in show business— an example being that on the very first show he booked Dean Martin and Jerry Lewis. It is often mistakenly believed that Elvis Presley made his television debut on Sullivan's program, but the fact was, Elvis made appearances on three other television programs before the *Ed Sullivan Show*. Sullivan learned from his mistakes, and made sure he booked popular rock n' roll performers first if he possibly could, such as the Beatles in 1964.

Sullivan would credit his Catholic faith to his not being able to tolerate prejudice against black performers. He did not care what race you were—if you had talent, you would be treated fairly. Ed had a business executive thrown out of the Ed Sullivan Theater for questioning Ed on why he had so many black acts on his show. Sullivan had to be physically restrained from beating up an automotive dealer who asked Sullivan why he put his arm around Bill "Bojangles" Robinson at the end of Robinson's dance routine. In fact, when Robinson died flat broke, as his friend, Ed paid for his funeral. Another example of Sullivan's generosity was with welterweight and middleweight boxing legend Sugar Ray Robinson. When Sugar Ray faced financial difficulties, Sullivan booked him on his show so that Sugar Ray could perform his dancing technique.

As we point out in our Introduction, like all of us, Ed Sullivan had his negative traits also. Adjectives like vindictive, nasty, and insulting have been hurled at him, but in the end we feel justified in including him in this volume.

The Ed Sullivan Show's 23-year-long run concluded in 1971. In 1974, Ed Sullivan died from esophageal cancer. His crypt is located in Ferncliff Cemetery.

FUN FACT

We have already indicated that Elvis Presley's first television appearance was not on the Ed Sullivan Show. It is mistakenly believed that Sullivan had Elvis filmed only from the waist up. Elvis appeared on Ed's program a total of three times, and the limited view was only on Elvis's third and final appearance.

CEMETERY

Ferncliff Cemetery
280 Sector Road
Hartsdale, NY 10530
Tel: 914-693-4700
Hours: Mon.–Sun., 9 AM–4 PM

DIRECTIONS TO THE CRYPT

Enter Ferncliff at 280 Sector Road and bear left; park toward the left side of the main mausoleum. Enter through the bronze doors and turn left at the first corner, then at the next corner turn right, followed by a left turn, then make another left, and finally make a right turn. Then proceed to the end of the last hall. Ed's crypt is near an elevator.

EDDIE RABBITT
Country Singer/Songwriter
Born: November 27, 1941
Died: May 7, 1998

We had always been fans of Eddie Rabbitt, and in preparing and researching him we got reacquainted with his music, much of which we hadn't heard in some time. Not only did his music bring back fond memories, but we were taken aback at the number of hits that were in his recording catalogue. However, that aside, why do we consider Eddie Rabbitt a hero and a worthy subject for this book? Rabbitt experienced a huge personal loss when his toddler son, Timothy, died just before his second birthday as the result of a rare children's disease affecting the liver. Eddie then became very involved in raising monies for organizations that helped sick children. We will get more into this below.

Eddie Rabbitt was born to Irish immigrant parents in Brooklyn, New York and raised in nearby East Orange, New Jersey. His earliest musical influence was his father, who played the fiddle and the accordion. Rabbitt gravitated to a love of country music, proclaiming himself "a walking encyclopedia" on that genre of music. Rabbitt said that the roots of country music came from Irish music but that he incorporated some minor guitar chords to achieve a "mystical" feel to his music.

Eddie Rabbitt recorded a few songs for Columbia Records and 20th Century Records in the mid-1960s. He moved to Nashville in 1968 and

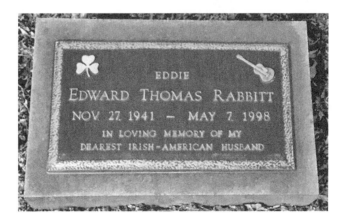

Eddie Rabbitt's Grave

made a name for himself as a songwriter when he wrote "Kentucky Rain," which was recorded by Elvis Presley and became a gold record for the King in 1970. In 1974, Eddie wrote "Pure Love," which was sung by Ronnie Milsap and went to number one on the country charts.

All this led in 1975 to a recording contract for Eddie Rabbitt with Elektra Records. "You Get to Me," his first single release on Electra, made the top 40. Eddie's next two singles were both top-20. In 1976, "Drinkin' My Baby Off My Mind" was his first number one country hit that he sang himself. In 1977, he was voted "Top New Male Vocalist of the Year" by the Academy of Country Music.

From 1976 through 1989, he charted over twenty-five number one country hits. Some were crossover hits such as "I Love a Rainy Night," which went to number one on the Hot 100 Chart. Other hits include "Every Which Way But Loose" from the Clint Eastwood movie of the same name, "Drivin' My Life Away," and "You and I" (a duet with Crystal Gayle).

After the death of Eddie's son, he spent much of his time fundraising for the American Council on Transplantation, as well as being a celebrity spokesman for the Muscular Dystrophy Association, the United Cerebral Palsy Association, the Special Olympics and Easter Seals. He raised millions of dollars for each of these causes.

Tragically, Eddie Rabbitt was diagnosed with lung cancer in 1997 and died the following year. He was just 56 years of age.

CEMETERY

Calvary Cemetery
1001 Lebanon Pike
Nashville, TN 37210
Tel: 605-256-4590
Hours: Mon.–Thurs., 7:30 AM–4 PM; Fri., 7:30 AM–3 PM; closed Saturday, Sunday and major holidays

DIRECTIONS TO GRAVE

Enter from Lebanon Pike gate and make a right turn. Stay on road, pass the utility shed, which is on the left. Further down the road that turns left, go approximately two to three car lengths. Then stop and look left approximately 75 to 100 feet for a large cross, Rabbitt's gravesite.

FANNY BRICE
Radio Personality/Entertainer
Born: October 29, 1891
Died: May 29, 1951

As one of us was involved in the radio industry for many years, when we discovered that comedienne Fanny Brice's most lasting legacy was her years as a radio personality, we had to include her in this volume. In our opinion, she was a trailblazer as well as an inspiration for other female comediennes. In the days before television, families would gather in the evening and listen to popular radio programs such as Fanny Brice's *Baby Snooks Show*. The character of Baby Snooks was that of a troublesome young girl known for throwing tantrums at her long-suffering father. The character of "Baby Snooks" was first introduced on radio in 1933, and Brice played this role until her death in 1951.

Fanny Brice's Grave

Fanny Brice was born as Fania Borach into a Jewish family in New York City. Brice's career would span several genres, as she appeared on stage, radio, film and television.

After dropping out of high school, Brice began working in a burlesque revue in 1908. In 1910, she caught the eye of Florenz Ziegfeld, the creator of the famed Ziegfeld Follies. The Ziegfeld Follies was an elaborate Broadway theater variety show that highlighted the diverse talents of many entertainers. The show became famous for its beautiful chorus girls, who were elaborately costumed for their dance numbers. Brice earned as much $3,000 per week when she became established with the follies. Her comedy act at the follies included humorous satires that won her a devoted following. Brice introduced the song "My Man" in 1921, which would become her signature song. She portrayed herself in follies movies such as 1936's *The Great Ziegfeld* and 1946's *Ziegfeld Follies*.

Brice was known for never rehearsing her "Baby Snooks" radio show, as she felt rehearsing ruined the spontaneity she wanted to achieve. She was so meticulous about the show that she even wore toddler-styled clothes so she could be in character. She would morph into Baby Snooks while performing on the air and in front of the small radio studio audience. Brice portrayed Baby Snooks only once on television in 1951.

Brice died of a cerebral hemorrhage at the age of 59 in 1951. Brice's passing occurred on a day her radio program was scheduled to air. The program that night was turned into a sort of eulogy for her and for her famous character.

FUN FACT

Barbara Streisand portrayed Fanny Brice on the Broadway stage in 1964 in the musical *Funny Girl*. Streisand also starred in and received an Oscar for her role in the 1968 film adaptation of *Funny Girl*.

CEMETERY

Westwood Village Memorial Park
1218 Glendon Avenue
Westwood, CA 90024
Tel.: 310-474-1579
Hours: 8 AM–5 PM

DIRECTIONS TO GRAVE

Proceed through entrance on Glendon Avenue and make a right. When you come to office building, make a left. Take all the way to the end, past a line of graves on the right of Jack Lemmon and Carroll O'Conner. On the right, you will see a path where there is a columbarium that is gated. This is the Garden of Serenity. Through the gate at the top of the niche in a small pillar, you will see the resting place of Fanny Brice's ashes.

PAUL NEWMAN
Actor/Director
Born: January 26, 1925
Died: September 26, 2008

The name of Academy Award–winning actor Paul Newman hardly needs an introduction. His roles in movies such as 1969's *Butch Cassidy and the Sundance Kid* and 1973's *The Sting* are classic movies due in large part to the strong performances of Newman. Even when Newman played complex roles, no one can deny he projected a likable quality. An example would be our personal favorite film of Paul Newman, which is 1961's *The Hustler*. Additionally, his philanthropic legacy continues to this day as his Newman's Own Foundation continues to give 100 percent of its profits to charities, with an emphasis on children's foundations. Now that's a terrific legacy!

Newman was born in Shaker Heights, Ohio. His father was Jewish and of Polish and Hungarian extraction, and his mother was a Christian Scientist and of Slovakian heritage. He served in the U.S. Navy during World War II. After the war, he received his BA degree in drama and economics from Kenyon College. He attended the Yale School for Drama before he studied under Lee Strasberg at the Actors Studio. In 1952, he made his first credited role on a television series called *Tales of Tomorrow*. In 1953 Newman made his Broadway debut in the play *Picnic*, and it was here that he first met his future second wife, Joanne Woodward, whom he would marry in 1958. (Newman's first marriage to actress Jacqueline Witte ended in divorce shortly before he married Woodward.)

The first movie role that earned him critical acclaim was in 1956 for *Somebody Up There Likes Me*, where Newman portrayed middleweight boxing champion Rocky Graziano. Another strong performance in *Cat on a Hot Tin Roof*, opposite Elizabeth Taylor and Burl Ives, earned more accolades from fans and critics as Paul Newman received his first Oscar nomination.

Newman in the 1960s continued his strong performances in films such as 1961's aforementioned *The Hustler*, 1963's *Hud*, and 1967's *Cool Hand Luke*. All three of these roles earned Newman additional Oscar nominations. But it wouldn't be until 1986 that Newman would finally win an Academy Award for best actor for reprising his role in *The Hustler* as Fast Eddie Felson opposite Tom Cruise.

It was in the early 1980s that Paul Newman conceived the idea of selling his bottled salad dressing. He and his friend, writer A. E. Hotchner, founded the company Newman's Own. Eventually the product lines included cookies and sauces. This company has to date donated hundreds of million dollars to various charities worldwide.

Paul Newman died in 2008 as the result of lung cancer. He was cremated with his ashes given to his wife.

FUN FACT

In 1985, Paul Newman was awarded an honorary Academy Award, which he accepted by saying in his speech, "I am especially grateful that this did not come wrapped in a gift certificate to Forest Lawn."

BOB HOPE
Comedian/Entertainer
Born: May 27, 1903
Died: July 27, 2003

Bob Hope will certainly go down as being one of America's greatest comedians. His rapid-fire one liners and comedic timing influence performers even today. Unlike many of the stand-up comedians of today, many of Hope's routines were self-deprecating. However, the main reason for Hope's inclusion in this volume was his dedication to

Bob Hope's Grave

entertaining America's troops on his USO tours, of which he did a total of 57 between 1941 and 1991.

Hope was born in London, England but was raised mostly here in the U.S., as his parents emigrated (more specifically) to Cleveland, Ohio. Hope in his late teens gravitated to show business. He started out in vaudeville. In 1933, he landed his first major role in the Broadway production of the musical *Roberta*. Bob then proceeded to become a very popular star on the NBC Radio Network with *The Bob Hope Show*.

It wasn't long before Hollywood began to cast Hope in successful comedies starting with *The Big Broadcast* of 1938. This movie featured what would become Bob Hope's signature song, "Thanks for the Memory." Probably his most popular movies were the *Road* pictures, such as *Road to Singapore* and *Road to Morocco*, with Bing Crosby. In these films, Hope and Crosby usually competed for the affections of Dorothy Lamour. Our favorite is the *Road to Utopia*. The scene with Hope at a rough-looking bar, ordering a lemonade "in a dirty glass" always brings a smile to our faces.

After success in radio and in the movies, television was next on Hope's list of achievements. Under contract with NBC, through the decades Bob Hope starred in innumerable specials, at least two a year.

Like Bing Crosby's "White Christmas," Hope also had a Christmas song, "Silver Bells," and he would always sing it on his Christmas specials. This song came from the movie *The Lemon Drop Kid* in which Hope starred. His first TV special was broadcast in April 1950 and his last special was done in November 1996. Do you think NBC got its money's worth? In addition to his TV specials, to this day, Bob Hope holds the record for hosting 19 consecutive Academy Awards.

Bob Hope, as it turns out, had quite a reputation as a womanizer. However, to his legion of fans it wasn't important enough to stop them from going to his movies and watching those memorable television specials. He remained married to his wife Dolores, whom he met when she co-starred with Bob in the Broadway show of *Roberta*, for over 60 years.

Bob Hope died at the age of 100 in July of 2003. The immediate cause was pneumonia. Dolores Hope passed away in 2011 at the age of 102.

FUN FACTS

It was reported that when Bob Hope was on his deathbed, he was asked where he would like to be buried. Hope responded by saying, "Surprise me."

In 1977, Hope and Crosby had plans to make one more *Road* picture with the tentative title of *Road to the Fountain of Youth*. Unfortunately, Bing Crosby died in October of that year, and plans were, for obvious reasons, dropped.

CEMETERY

San Fernando Mission Catholic Cemetery
11160 Stranwood Avenue
Mission Hills, CA 91345
Tel.: 818-361-7387
Hours: 8 AM to 5 PM

DIRECTIONS TO GRAVE

You must go to the church on the cemetery property. There you will find a gift shop where you pay a modest donation for access to Bob's gravesite behind the church.

MARIAN ANDERSON
Contralto Singer
Born: February 27, 1897
Died: April 8, 1993

Marian Anderson was a black American contralto who sang a wide range of music from opera to spirituals. She was an important figure for African Americans in their fight against racial prejudice in the 20th century.

Born in Philadelphia in the late 19th century to a family of modest means, Marian's father, John, sold ice as well as coal, and her mother, Annie, was a nanny. As a child, she began singing in the choir at the Union Baptist Church. At the age of six she earned a quarter for a benefit concert, and by her teens was earning $5 for a show.

Due to racial prejudice, Anderson struggled as a young adult to secure enough bookings to earn a living as a singer. Her big break came in winning a singing competition with the New York Philharmonic. As a result, she performed in a concert with the orchestra in August 1925. Despite adversity, she persevered and secured performances in a variety of venues in the United States. Her first concert at Carnegie Hall was in 1928.

In the summer of 1930, Anderson left for Scandinavia, where she met the Finnish pianist Kosti Vehanen, who became her vocal coach and accompanist. Through Vehanen she met the renowned Finnish composer Jean Sebelius. They became friends and formed a professional relationship, with Sibelius composing songs for Anderson. Her career began to thrive in Europe, and "Marian Fever" took hold on the Continent. In a 1935 concert in Salzberg, Arturo Toscanini told her she had a voice "heard once in a hundred years." She also toured Russia and Eastern Europe to much acclaim.

Segregation, however, still continued to plague her in her native land.

Constitution Hall in Washington, DC refused her a platform for a concert on Easter Sunday on April 9, 1939. The Daughters of the American Revolution (DAR) was the group behind the denial because Constitution Hall was segregated. This caused incredible backlash by the NAACP that reached all the way to the White House.

Marian Anderson's Grave

First Lady Eleanor Roosevelt, a DAR member resigned from this organization in protest. At her instigation, she got FDR to have his Secretary of the Interior, Harold Ickes, arrange an open-air concert on the steps of the Lincoln Memorial.

On that crisp Easter Sunday, Marian, wrapped in a fur coat against the chill, launched into "America" before the assembled crowd of 75,000. In his introduction, Ickes said, "Genius draws no color lines." She also sang "Ave Maria," "Nobody Knows the Trouble I've Seen" and "Gospel Train," among many others.

Her Lincoln Memorial performance allowed her to break national barriers. She entertained troops in World War II and the Korean War. In 1955, she became the first black performer at the Metropolitan Opera. The following year her autobiography, *My Lord What a Morning*, was published and became a best seller.

Her global stature reached new heights in 1957 when she was appointed good-will ambassador for the state department, giving concerts worldwide. She also sang at the presidential inaugurations of both Dwight Eisenhower and John Kennedy. In 1963 she sang at the March on Washington for jobs and freedom. After her farewell concert at Carnegie Hall in 1965, she retired from active singing. Many honors were accorded to her including the Congressional Gold Medal, Kennedy Center Honors and a Grammy Award for lifetime achievement.

In her retirement, she spent time out of the public eye with her husband, Orpheus "King" Fisher, an architect, on their 100-acre farm "Marianna" in Danbury, Connecticut. Sometimes she would sing when the town lit up their Christmas tree. In 1992, six years after the death of her husband, she relocated to the home of her nephew in Portland, Oregon. It was there that she died the following year of congestive heart failure at the age of 96. She had lived her life marching to her own drum. She was interred in Eden Cemetery in Collingdale, Pennsylvania next to her mother and sister.

FUN FACTS

There is an astonishing legacy of 26 recordings of Marian singing all genres of music under different record labels. The most dominant is RCA Victor.

In 1960, Maria Cole, the wife of singer Nat King Cole, asked Marian to affix her signature to a statement along with other prominent black leaders endorsing John Kennedy for president. Anderson refused, stating that it was "not possible" to have her name "identified with any particular party."

CEMETERY

Eden Cemetery
1434 Springfield Road
Collingdale, PA 19023
Tel.: 610-583-8737
Hours: Daily, 8 AM–4:30 PM

DIRECTIONS TO GRAVE

Upon entering the cemetery gates, turn left. About 100 feet in on the left will be Marian Anderson's grave marker with her last name.

WILLIAM S. PALEY

Broadcast Pioneer
Born: September 28, 1901
Died: October 26, 1990

William Paley was a seminal figure in the history of broadcasting. He built the Columbia Broadcasting System (CBS) from a small radio network and transitioned it into the CBS Television Network, one of the top broadcast entities in the United States in the mid-twentieth century.

William Samuel Paley was born in Chicago, Illinois to Goldie and Samuel Paley. His family was Jewish and his father, a Ukrainian immigrant, ran a cigar company. The cigar company became very suc-

William S. Paley's Grave

cessful and Samuel Paley became a millionaire early on. Academically, William matriculated at Western Military Academy in Alton, Illinois and received his college degree from the Wharton School of Business at the University of Pennsylvania. His family assumed William would assume the running of the cigar business.

In 1927, Samuel Paley and some business partners purchased a struggling Philadelphia-based radio network of 16 stations called the Columbia Phonographic Broadcasting System. The initial thought was to use the network to promote the cigar business, whose sales under William had doubled. In 1928, the family had a majority interest in the network.

William Paley quickly grasped the earnings potential of radio, and within a decade the network had 114 affiliate stations. He recognized that good programming was the key to bringing profits to the network and affiliate stations. Initially, individual stations bought programming from the network and were considered the network's clients. Paley changed that business model by developing successful and lucrative network programming at a nominal cost to affiliates, and by viewing advertisers as the most significant component of the business model. The advertisers were the network's primary clients, and with wider distribution and the corresponding audience, he was able to charge more for ad time. During part of the broadcast day, affiliates could carry local programming that they could sell locally.

During World War II, Paley recognized American's desire for news coverage and built the news division into a powerful force to complement the dominant entertainment arm of the network. During the war, he was Director of Radio Operations of the Psychological Warfare Branch in the Office of War Information in London with the rank of colonel. It was there he met and befriended Edward R. Murrow, CBS's head of European news, who expanded the war coverage with a stable of correspondents known as Murrow Boys.

In 1946, Paley promoted Frank Stanton to president of CBS. Paley expanded into TV and surpassed NBC in coverage and profits. His friendship with Edward Murrow suffered in the 1950s due to the hard-hitting tone of Murrow's show, *See It Now*, which covered controversial topics. Paley worried about the potential of lost revenue, which came to pass when Alcoa withdrew its sponsorship, causing the eventual termination of the weekly series after 1958. In the 1960s CBS enjoyed a golden era of television programming, which earned it the nickname of "the Tiffany Network." Shows such as *Gunsmoke*, *Gilligan's Island*, *The Beverly Hillbillies*, *The Ed Sullivan Show*, and *The Wild Wild West* secured CBS's dominance of ratings and profits. The *CBS Evening News* became the number one news source with Walter Cronkite at the helm. Cronkite was dubbed "America's Most Trusted Man."

Paley also ventured outside of broadcasting with the acquisition of CBS Records in 1939, which introduced the 33-1/3 RPM record, capable of playing 20 minutes of music on each side, in 1948. This type of record continued to be the standard into the 1970s. He also purchased an interest in the New York Yankees baseball team in 1964. They were a mediocre team at the time and failed to make the post season in the nine years CBS owned them. Paley sold the team to Cleveland shipbuilder George Steinbrenner in 1973.

Paley was known as a ladies' man. His first wife was Dorothy Hearst, whom he married in May 1932. They became estranged because of his infidelities and divorced on July 24, 1947. His second wife could be called the love of his life. Socialite and fashion icon Barbara "Babe" Cushing Mortimer wed Paley four days after his divorce. They had two children, William and Kate. This marriage survived until Babe's death due to cancer in 1978.

Philanthropically, he was active in the Museum of Modern Art in New York City, becoming chairman of the Board of Trustees. He helped acquire six Picassos from the Gertrude Stein collection for the museum. In an homage to his broadcasting roots, he founded the Museum of Broadcasting in New York City. The name was later changed to the Museum of Television and Radio. Today, it is known as the Paley Center for Media.

Over the years, as he wound down his career at CBS, he sold portions of the family's stock holdings. At the time of his death he owned less than nine percent of outstanding stock in CBS.

Paley succumbed to kidney failure on October 26, 1990, at the age of 89. He was interred next to his beloved wife at the Memorial Cemetery of St. John's Episcopal Church.

FUN FACTS

To give New Yorkers a breath of fresh air, Paley constructed a vest-sized park with a waterfall on 53rd Street off Fifth Avenue on the site of the old Stork Club. It is known as Paley Park and is an oasis in the middle of Manhattan.

As previously noted, as an eligible bachelor Paley was noted for squiring women around town. He was on the list of the ten most eligible bachelors by *Cosmopolitan Magazine* in 1985 at the age of 84. The irony of the octogenarian Paley being on the list was the inspiration for David Letterman's Top Ten Lists.

CEMETERY

St. John's Episcopal Cemetery
1704 Route 25A
Syosset, New York 11791
Tel.: 516-692-6748
Hours: Dawn to dusk, 7 days a week

DIRECTIONS TO GRAVE

Navigating this bucolic country cemetery can be challenging since there are no road signs or section markers. It is best to get a map at the cemetery office to try and guide you. On the map, Paley will be in section 8. From the office, turn right and take the road down. It meanders a bit but stay on that road. About a quarter of a mile down before it splits into a fork, you will find the Paley plot on the left. It is a small alcove surrounded by trees and has a bench in it.

MARIE TORRE
Journalist/TV News Anchor
Born: June 17, 1924
Died: January 3, 1997

Adherence to principle is why we have included Marie Torre in this second volume of *Grave Trippers*. More specifically, adherence to the principle of free speech and to the United States Constitution, as stipulated in the First Amendment.

Marie Torre was born in Brooklyn, New York, and from the time she was a young girl she knew that she wanted to be a journalist. After graduating from high school in 1942, she managed to get a job as a "copy boy" at the *New York World Telegram & Sun* while attending New York University at night. She worked her way up quickly to become a reporter at the newspaper. In 1952, she joined the staff of the *New York Herald Tribune*. She made history when she became one of the youngest feature editors in the city. She wrote a widely-read syndicated entertainment column that featured many of the big names of the day, such as Marilyn Monroe, Marlon Brando, Eleanor Roosevelt, Montgomery Clift, Frank Sinatra and former President Harry Truman.

In 1948 she married Harold Friedman, a television producer. In 1957, Marie gave birth to her son, Adam, and in 1958, daughter Roma arrived.

Marie made national headlines after she wrote an article in January 1957 concerning Judy Garland and CBS. On a tip from a confidential CBS source, Marie wrote that Judy was being difficult with the network after she had been contracted to perform in a live concert special. Marie quoted the source as saying he believed the problem was that "Judy thinks she looks terribly fat." Judy had no comment initially, but three months later she sued CBS for nearly $1,400,000.

At a pretrial hearing, Marie Torre was ordered by the court to name the source for her story. Marie refused, stating that to compel her to name her source was a violation of the First Amendment, guaranteeing freedom of the press. Various news organizations rallied to her defense when a judge sentenced her to thirty days in jail for contempt of court. Marie's lawyers appealed the charge all the way to the U.S. Supreme Court, but the Justices refused to take the case.

Marie Torre's Grave

The contempt charge was upheld by Judge Potter Stewart, who later joined the Supreme Court. The sentence was reduced to ten days, but Marie was warned that she would be sent back to jail again and again as long as she refused to name her source. The case received world-wide attention, and at the end of the ten days, the judge withdrew his threats to send Marie back to jail. Years later, Judy Garland dropped the charges.

With her national notoriety, Marie Torre was offered the opportunity of a lifetime. In 1962, Pittsburgh's KDKA-TV hired her to become the nation's first female news anchor, where she would be "covering news like a man." Over the years her popularity grew, and she was given her own program to host. The highly-rated *Marie Torre Show* lasted till 1976, when Marie decided to return to her hometown, New York City, following the death of her beloved husband. In her 14 years in Pittsburgh, she made an indelible mark covering hard news and interviewing a wide-ranging list of newsmakers including President Lyndon Baines Johnson, Joan Crawford, Muhammed Ali, Coretta Scott King and Alabama Governor George Wallace. Her career thrived back in New York as well, where she went on to write and produce many television specials, earning numerous awards including three Emmys.

Marie Torre passed away on January 3, 1997 as the result of lung cancer. Her precedent-setting case helped scores of future reporters avoid jail.

FUN FACT

Despite citing her for contempt, the sentencing judge praised Marie Torre for her principled stand, calling her "the Joan of Arc of her profession."

CEMETERY

Jefferson Memorial Park
401 Curry Hollow Road
Pleasant Hills, PA 15236
Tel.: 412-655-4500
Hours: Mon.–Fri., 9 AM–5 PM; Sat., 9 AM–4:30 PM; closed on Sunday

DIRECTIONS TO GRAVE

Enter main cemetery entrance and turn left to the office. In front of the office, go straight ahead past the mall. At the end of the mall, turn right to the Jefferson Mausoleum. Follow it around, turning left as you do. Take your first right. Follow straight to the first intersection. Garden of Peace will be on the left. Turn left, then make first right at Garden of Peace. About 50 feet in the second row on a slight hill in front of a tree will be Marie Torre's resting place, marked Lot 70.

CHAPTER SEVEN
Clergy

ARCHBISHOP FULTON SHEEN
Clergy
Born: May 8, 1895
Died: December 9, 1979

Archbishop Fulton J. Sheen was born in Peoria, Illinois to a family of Irish descent. What made us want to include Bishop Sheen in our book about heroes? Because we feel that he walked the walk and talked the talk. Throughout his life, Sheen donated to the poor and needy almost all of the money that he earned and fundraised through his books and television appearances. He didn't care if you were rich or poor; Sheen would spend as much time as necessary if he thought you needed having your soul saved. Sheen not only brought back into the Church wayward Catholics, he is also credited with converting atheists, agnostics, communists, Protestants and Jews to the Catholic faith.

Ordained a priest in 1919, Sheen earned two doctorates, first from the Catholic University of Louvain in Belgium in 1923, and second from the University of Saint Thomas Aquinas in Rome, Italy in 1924. From 1926 through 1950, then Father and later Monsignor Sheen taught philosophy at the Catholic University of America. Students who took Sheen's courses would say that it was rare for any student to ask questions. So keen was Sheen's intellect and logic as well as his mesmerizing persona and delivery that students in his classes would just sit and listen until the class ended.

As Sheen's reputation for oratory grew, in 1928 he was invited to speak on the radio for NBC's *Catholic Hour*. With his amazing knowledge and distinct voice, Sheen quickly became popular nationwide, and he found himself invited to speaking engagements across America. Sheen's radio career lasted over 20 years, ending in 1950. A new medium would soon replace radio and would provide an even larger audience for Sheen to reach. That medium would be television.

In 1951 Sheen was consecrated a bishop, and in 1952 was offered the opportunity to appear on television. The program was titled *Life Is Worth Living*, and it ran for five years, ending in 1957. The program, which consisted of Bishop Sheen just speaking into a camera and using a blackboard to write down his major points of discussion, was a major hit. Bishop Sheen would win two Emmys for the half-hour show. At its peak, *Life Is Worth Living* had an estimated 30 million viewers each week.

In 1958, Sheen was made national director for the Society for the Propagation of the Faith, which was an organization dedicated to assisting missionary priests, nuns, and brothers. Bishop Sheen would return to television again in 1961 for *The Fulton Sheen Program*, which in essence used the same format as *Life Is Worth Living*. Most of these programs were in color. People were always amazed at not only the depth of Bishop Sheen's knowledge and the variety of topics he would cover during the program, but also his ability to end each show precisely when he needed to. He accomplished this feat by having a very large clock that indicated how much time he had left to speak. When it indicated that he had two minutes left, Bishop Sheen knew that he would have to quickly wrap it up. He concluded each show as he concluded every sermon he ever gave, with "God Love You."

Archbishop Fulton Sheen's Tomb

In October 1966, Francis Cardinal Spellman had Bishop Sheen reassigned to Rochester, New York from New York City as a result of a falling-out between the two. Though they were never publicly critical of one another, it was no secret that Sheen and Spellman did not see eye to eye on everything. But whatever the specific reason, Cardinal Spellman was Bishop Sheen's superior, and Sheen was in no position to refuse the reassignment.

As it turned out, Sheen's exile from New York City was relatively brief, as he resigned from his Rochester administrative duties in October 1969. In the previous month Sheen had marked his 50th anniversary of ordination as a priest.

Sheen spent the rest of his life primarily concentrating on writing articles, columns and books. On December 9, 1979, Archbishop Fulton J. Sheen died due to complications from open heart surgery.

Since the late 1990s, the formal process for canonization of Bishop Sheen has been underway. In order to be a declared a saint in the Catholic Church, at least two miracles must be attributed to the individual in question. As of the time of the writing of this book, the Church has recognized one miracle that has been attributed to him.

FUN FACTS

Whenever someone would ask Bishop Sheen to what he attributed his high television ratings, Sheen would respond by saying he always had better writers than his competition. And those writers were Matthew, Mark, Luke and John.

In preparing for his weekly television broadcast, Bishop Sheen would practice it in both French and Italian. Sheen felt that doing so forced him to better understand and communicate his chosen topic to his audience.

TOMB LOCATION

The Cathedral of Saint Mary of the Immaculate Conception
607 NE Madison Avenue
Peoria, Illinois 61603
Tel: 309-671-1550
Hours: Mon.–Fri., 12 PM–2 PM; Sat., 3 PM–5 PM; Sun., 10 AM–1 PM

CARDINAL EDWARD EGAN
Clergyman
Born: April 2, 1932
Died: March 5, 2015

When the names of leading American clergy of the past two centuries come up, one name that isn't mentioned often is that of Cardinal Edward Michael Egan. Leading the apostolates of Bridgeport, Connecticut and New York City, he was nevertheless responsible for an astounding number of accomplishments.

Born in Oak Park, a suburb of Chicago, Egan contracted polio as a child and was bedridden for two years. He overcame his handicap and was home schooled by his mother. He took up piano and was considered an excellent pianist, but he viewed it rather as a lifetime avocation. His mind was set on becoming a priest.

Academically proficient, Egan entered St. Mary of the Lake Seminary and compiled a remarkable 98.6 grade point average. Cardinal Samuel Stritch of Chicago said he should continue his studies at the Pontifical North American College in Rome. He was ordained a priest while there on December 15, 1957. In July 1958 he was awarded a degree in sacred theology, magna cum laude.

After arriving as curate in Chicago, he was selected by Cardinal Albert Meyer to be his secretary because of his Vatican contacts and the fact he spoke fluent Italian. In August 1960 he was appointed to the faculty of the North American College in Rome. While there, he managed to earn a doctorate in canon law.

In June 1965, the then new archbishop of Chicago, John Cody, recalled Egan to be his secretary. In 1971, Egan left and became a judge on the Tribunal of the Sacred Roman Rota. In addition, he taught canon law at the Gregorian University and civil and criminal procedure at the Studio Rotale in Rome. Egan was fortunate to be appointed one of the five canonists to work on the new Code of Canon Law. This gave him a tremendous amount of exposure to Pope John Paul II, who had a great interest in the project.

In April 1985, Egan was appointed Auxiliary Bishop of New York under Cardinal John O'Connor. He was the vicar for education during his tenure.

In December 1988, Egan was appointed Bishop of Bridgeport, Connecticut. He inherited a financial mess. He copiously fundraised, tapping Jack Welch of GE, who became a major donor. He stabilized finances and started the Inner City Foundation for Charity and Education, and additionally built four new parishes and two assisted living facilities, and helped various charitable agencies. Egan also handled the thorny situation of nine priests accused of sexual harassment, which had occurred prior to his administration. He would not tolerate any priest who was guilty of these offenses.

Upon the death of Cardinal O'Connor in early 2000, he inherited yet another fiscal and administrative quagmire as the new arch-

Cardinal Edward Egan's Tomb

bishop. The archdiocese of New York had a projected $20+ million deficit, the common fund, which comprised general revenues for the archdiocese, was $47 million short and the hospital system owed creditors $300 million.

Archbishop Egan then set about doing what he did best: fundraise. He always said, "You know I'm a beggar." He also visited as many parishes as he could. By his second year, he had been to over 150 parishes, over a third of the total. In February 2001, Pope John Paul II elevated him to cardinal.

In 2001 when 9/11 occurred, he donned scrubs and started anointing the sick and dead at the scene after watching the second tower fall.

The priest/predator situation also reared its ugly head in New York, and Egan attacked it with the same zeal as he had done in Bridgeport. He had a zero-tolerance policy for priests who participated in sexual abuse, and they were removed from the ministry.

Cardinal Egan also had to make tough decisions on schools, parishes and aging facilities. With his financial acumen and key advisors, he set upon a path to correct these problems. With the churches alone, he found donors to renovate thirteen churches, expand six and start construction on seven more.

As in Bridgeport, he started the Inner-City Scholarship Fund and raised more than $140 million with an endowment of over $100 million. Egan even cultivated an atheist who gave the schools in excess of $45 million.

In the broadcast arena, he was responsible for the establishment of the Catholic Channel on Sirius Radio. Its programming runs 24/7.

The crown jewel of his reign as Archbishop of New York was the three-day visit of Pope Benedict XVI in April of 2008. With Pope Benedict, he visited ground zero and celebrated mass at Yankee Stadium.

Egan became the first Archbishop to retire, in 2009 at the age of 75. He decided to stay in New York and took up residence at the Chapel of the Sacred Hearts of Jesus and Mary. In retirement he did not slow down, and aided his successor, Timothy Dolan, filling in at events, fundraising and performing confirmations. Dolan praised Egan when they saw the pontiff together in the Vatican, and Egan said that he was happy to help while he could. Pope Benedict said, "Bravo, Eminenza, Bravo."

Cardinal Egan suffered a fatal heart attack after just finishing lunch at his residence on March 5, 2015. He was entombed beneath the main altar of St. Patrick's Cathedral in the crypt reserved for the Archbishops of New York.

FUN FACT

At a gathering of firefighters some years after 9/11, Cardinal Egan ran into then NYC Fire Commissioner Salvatore Cassano. Cassano related to Cardinal Egan that on 9/11 he was hit with falling debris and was lying on the ground semi-conscious when Egan came up to him, knelt and spoke in his ear, telling him that he was absolving him of his sins. The Cardinal was taken aback, stating that it was dark and how could Cassano have known it was the Cardinal? "Everyone knows your voice!" was Cassano's response. Cardinal Egan had a deep baritone voice that was easily recognizable and was often called "the voice of God."

ENTOMBMENT

St. Patrick's Cathedral is located on NYC's Fifth Avenue between 50th and 51st Street in Manhattan and is open from 8:30 AM to 6:30 PM. Behind the main altar, there are steps that lead down to the archbishop's crypts. The cathedral offers guided tours that include a visit to these crypts.

REVEREND NORMAN VINCENT PEALE
Clergyman
Born: March 31, 1898
Died: December 24, 1993

Norman Vincent Peale was a Protestant clergyman who in the 1950s and '60s introduced a new way of living through positive thought and actions. His runaway bestseller *The Power of Positive Thinking* paved the way for this approach, which was revolutionary at the time and attracted many converts to his new way of life. His influence extended to many notable people including Richard Nixon, with whom he had a very close friendship.

Peale was born in Bowersville, Ohio to Charles and Anna Peale. His father was a doctor turned Methodist minister. He graduated from Bellefontaine High School in a town of the same name in Ohio. He then earned a degree at Ohio Wesleyan University. He also attended the Boston School of Theology. His father convinced him to forego the formal preaching style of that school for one of simplicity in life and a search for the goodness in each human being. In 1922 he was also ordained a Methodist minister.

After his first pastoral assignment in Rhode Island, he landed in Brooklyn in 1924, where his leadership increased the membership twenty-fold in one year. In 1927, he received a call to take the pulpit at the University Methodist Church in Syracuse. It was there that he began to take his sermons to the new medium of radio. On June 20, 1930, he married Loretta Ruth Stafford.

In 1933, he made a switch in churches and also in religious denomination by going to assume the pulpit of the Marble Collegiate Church, which is located in New York City. This Protestant Reformed Church was founded in 1628 and the current structure was built in 1854. The initial congregation of two hundred grew to thousands due to his spirited sermons. He would remain there until 1984.

His first book, *The Art of Living* (1937), took its name from his popular radio program on the NBC Radio Network, reaching millions. Peale's radio program ran for an amazing 54 years. His first book was followed by *You Can Win* and *The Tough-Minded Optimist*. In 1945, Peale and his wife Ruth founded *Guideposts* magazine, which pre-

Reverend Norman Vincent Peale's Grave

sented inspirational stories in a non-denominational forum. In 1948, he wrote his first bestseller, *A Guide for Confident Living*, which brought religion to bear on personal problems. In 1952, he wrote the book he is best known for, *The Power of Positive Thinking*. Even today it is still hugely popular. Guidepost Publications in conjunction with the Peale Center is located in Pawling, New York.

Peale made several forays into the political sphere. His most famous was criticizing Adlai Stevenson as being unfit for the presidency because he was divorced. Stevenson returned fire by stating, "I find St. Paul appealing and St. Peale appalling." Eventually Stevenson apologized in person to Reverend Peale.

His influence was great, as evidenced by the fact that five U.S. presidents spoke well of him in the documentary about his life, *Positive Thinking: The Norman Vincent Peale Story*. Billy Graham stated that he didn't know anyone who had done more for the kingdom of God than Peale and his wife Ruth.

Peale died on Christmas Eve 1993 as a result of a stroke at the age of 95. He was interred in Quaker Hill Cemetery in Pawling, New York.

FUN FACTS

Peale, being close to the Nixon family, officiated at the wedding of Julie Nixon and David Eisenhower at Marble Collegiate Church in Manhattan on December 22, 1968.

Norman Vincent Peale's life story was portrayed in the movie *One Man's Way*. The actor who played Reverend Peale was Don Murray.

CEMETERY

Quaker Hill Cemetery
29 Church Road
Quaker Hill, New York 12564

Tel.: 845-855-5304
Hours: Dawn to dusk

DIRECTIONS TO GRAVE

Upon entering the cemetery, turn right onto the loop road. Bear right and you will pass a seating area and benches. Two-thirds of the way around the loop, his grave will be on the right. It is to the left of a pine tree.

CHAPTER EIGHT
Judiciary

LEARNED HAND
Federal Judge
Born: January 27, 1872
Died: August 18, 1961

Learned Hand was the chief justice of the U.S. Court of Appeals. His tenure spanned over 52 years. In his day, Hand was famous and was viewed as the 10th justice of the U.S. Supreme Court. Today, he is largely forgotten and unknown, even though many legal scholars consider him to have been a greater judge than all but a few who have sat on the highest U.S. court. He was known for his keen mind, philosophical skepticism and faith in the United States.

Learned was born in Albany, New York to Samuel and Lydia Hand. His father was an appellate lawyer who was distant to his son. Learned was not happy with his name and was beset with anxieties, self-doubt and insecurities throughout his life. At the age of seven, after two years at a small primary school, he transferred to the Albany Academy. Learned Hand finished at the top of his class and was accepted into Harvard College in 1889.

At Harvard he was known for his intense and studious ways that didn't allow him to socialize much. The fruits of his labor got him elected into Phi Beta Kappa, an elite scholarly society. Plagued by constant self-doubt, however, he was drawn into the study of law at Harvard Law School. At Harvard Law he was taught the nuances pertaining to evidence and constitutional law by Professor Bradley Thayer. Thayer became a major influence in Hand's jurisprudence, emphasizing the law's historical and human dimensions rather than its certainties and extremes.

Hand earned his bachelor's of laws degree in 1896 at the age of 24. He worked at his uncle's law firm in Albany but was displeased with court work. He found himself researching and writing briefs instead of the appellate work that he preferred. Despite entreaties by his family to remain in Albany, he began to seek law positions in New York City.

At the age of 30 Hand met his future wife, Frances Fincke, while on vacation at a Quebec resort. After a courtship of nearly two years, they were married, and their union produced three daughters. Eventually they rented a summer home in Cornish, New Hampshire, an

Learned Hand's Grave

artist colony with a stimulating social scene. However, since it was a nine-hour train ride and he was toiling at a Manhattan law firm, he could join his family only for vacations. Still unhappy with the practice of law, Hand lobbied for a potential new federal judgeship in the U.S. District Court for the Southern District of New York. With the help of his influential friend Charles Burlingham, he gained the backing of Attorney General George Wickersham, who urged President Taft to appoint him to the federal bench. One of the youngest federal justices ever appointed, Hand took the office at the age of 37 in April 1909.

As a judge he dealt with the field of common law including torts, contracts and copywrites. He also affirmed the right of free speech in his decision U.S. versus Kennerly in 1913. Hand became active politically, writing articles for the *New Republic* on social reform and judicial power.

In 1917 he rendered his most memorable decision in *Masses vs. Patten*. The Espionage Act forbade any criticism of the war effort in World War I. The Postmaster General of New York refused to deliver the first issue of *The Masses Magazine*, deeming it revolutionary since it supported those who refused to serve in World War I. Hand felt that the text did not tell readers to violate the law, stating that freedom of speech should be protected. The *Masses* decision, however, did not help Hand politically and did not aid him in seeking a promotion.

With the Republicans attaining power in 1921, things began to change. In 1923 Supreme Court Justice Oliver Wendell Holmes stated that he wanted Hand on the Supreme Court. Based on Hand's newfound stature, President Calvin Coolidge in 1924 appointed him to a higher federal court, the 2nd Circuit. A staunch Republican, he supported Herbert Hoover twice for president. Despite a hard lobbying effort by Hand's friend Felix Frankfurter, then a Harvard Law School professor, for political reasons Hoover appointed Charles Evans Hughes to the high court.

Hand increasingly became an acknowledged leader on New Deal statutes under FDR. In the 1935 case of *U.S. versus Schechter*, he ruled against the New Deal unfair trade practices. The U.S. Supreme Court later affirmed his ruling. In 1939, he became his court's chief judge. Until then largely unknown to the public, Hand's short speech to a million and a half people in Central Park in May 1944 was widely circulated by the *New York Times*, the *New Yorker*, *Life* and *Reader's Digest*. A key section of the speech was his definition of the "spirit of liberty." Hand stated, "The spirit of liberty is the spirit that is not too sure that it is right. The spirit of liberty is the spirit that seeks to understand the minds of other men and women, the spirit of liberty is the spirit that weighs its interest alongside theirs without bias." By his 75th birthday in 1947, he was celebrated and called "now unquestionably the first among American judges" by his former sponsor, Charles Burlingham.

At the onset of the Cold War, Hand was distrustful of Stalinism while also opposing McCarthy's anti-communist campaign. In 1951, he wrote an unscripted speech published by the *Washington Post* condemning McCarthyism. That same year he retired from "regular active service," assuming senior status but still carrying a heavy case load. In 1952, he published the book *The Spirit of Liberty*, a collection of his speeches that surpassed his expectations and made him more well known.

In his last decade, he was in generally good health and enjoyed life in Cornish, New Hampshire with his wife Frances. In 1958, he started to experience back pain and difficulty in walking. By June 1961, he was in a wheelchair. The following month, he suffered a heart attack and died in August at St. Luke's Hospital in New York City. He was buried next to his wife in the family plot in Albany's Rural Cemetery.

FUN FACT

Hand was walking after dinner one summer evening on New York's 6th Avenue. The streets were full of young strolling couples and brightly dressed lone females. Hand turned to his companion and said, "Do you think there is as much mounting going on as there was when we were younger?" His companion pondered the question and answered in the affirmative. Hand replied, "So do I, so do I. Good!"

CEMETERY

Albany Rural Cemetery
3 Cemetery Avenue
Albany, New York 12204
Tel.: 518-463-7017
Hours: Mon.–Fri., 8:30 AM–4:30 PM; Sat., 8:30 AM–12 PM;
    closed on Sundays and holidays

DIRECTIONS TO GRAVE

Enter cemetery at the Cemetery Avenue entrance. The office will be on your left. On the right you will see Ridge Road. Follow Ridge Road three sections on the right, between sections eight and twelve. Make a right and go straight to section eleven. Hand is in the family plot approximately in the middle of the section near some trees. His plain upright marker, where he lies with his wife, is to the left of the small temple-style marker for his parents.

SAMUEL SEABURY

Judge/Criminal Prosecutor
Born: February 22, 1873
Died: May 7, 1958

Samuel Seabury was an honest and crusading judge who cleaned up the corruption in New York's Tammany Hall politics, which was on a level in the early 20th century that rivaled the misdeeds of Boss Tweed in the 19th century. His investigations from 1930 to '32 were known as the "Seabury Hearings." The cleanup that ensued helped paved the way for Fiorello LaGuardia to be elected mayor of New York City. Seabury was born in the rectory of the Church of the Annunciation

Samuel Seabury's Grave

on West 14th Street in the borough of Manhattan to William Jones Seabury and Alice Van Wyck. His father became professor of canon law at the General Theological Seminary. Seabury grew up in modest circumstances. He eventually attended William & Kellogg High School for boys and graduated in 1890. Seabury then graduated from New York Law School in 1893 and was admitted to the bar in 1894.

Aspiring to be a judge, he initially ran in 1899 on both the Independent Labor and Republican tickets for the New York City Court. He was defeated by the Tammany Hall candidate. In 1901 he tried again, this time as the candidate of the Citizens Union ticket, and was elected to a ten-year term. He became the youngest judge on the court at 28 years of age.

On June 6, 1900, he married Maud Richey, the daughter of an Episcopal minister and seminary professor. They had no children. In 1906, he ran for the New York Supreme Court on the Democratic and Independent League fusion ticket and was elected to a 14-year term. In 1914, he ran for the New York Court of Appeals, the state's highest court, and was the candidate of the Democratic, Progressive, Independence League and American tickets. He won again and was elected to a 14-year term.

In 1916, former President Theodore Roosevelt, who opposed sitting New York State Governor Charles Whitman, persuaded Seabury to run for governor. While Franklin Roosevelt supported Seabury, Theodore did not obtain the Progressive nomination for Seabury, and he lost a close election. Seabury then resumed his private law practice. He came out of the private sector in 1930 to become the lead investigator of the Hofstadter Committee, a joint New York City legislative committee that scrutinized rampant corruption in New York's municipal courts and the New York City Police Department. The investigations, which ended in 1932, called over 1,000 witnesses.

Their investigative technique, which was pioneered by Seabury's chief counsel, Isadore Kresel, relied on gathering incredible amounts of facts pertaining to the investigation including bank account documents, brokerage accounts, leases, title records and income tax returns. These were used to confront witnesses during questioning. Prior to this technique, investigative committees relied on interviews and public testimony from confessors to inform on decisions and outcomes of investigations.

The climax of the Seabury Hearings came when Samuel Seabury cross-examined New York City Mayor Jimmy Walker for two days on May 25–26, 1932. Seabury read damning evidence into the public record, and Walker admitted to receiving a gift of $246,000 in a stock brokerage account. On June 8, he sent a transcript of Walker's testimony to Governor Franklin Roosevelt and asked that he review the evidence and make a decision to remove or not remove Walker.

On August 11, 1932, in the Executive Chamber, Governor Roosevelt began sitting as the judge in the removal proceedings against Mayor James Walker brought by Samuel Seabury and others. The tense proceedings ended on September 1, when Walker submitted his letter of resignation. He fled the city for Europe and ended up in Paris in exile. It should be noted that Seabury was never paid for these services. He did it out of his love for the city and good clean government.

In 1933, Seabury campaigned for the election of Fiorello LaGuardia as the Fusion candidate for New York City mayor, as he did all three times that LaGuardia was elected. La Guardia compiled a remarkable record in his 12 years cleaning up the mess he had inherited from Walker.

Seabury retired in 1950 to his home in East Hampton, New York. After suffering a hip fracture in 1955, he became an invalid, and his health problems were exacerbated by dementia. He died at Hand's Nursing Home in East Hampton on May 7, 1958. After a funeral at Trinity Church, he was interred next to his wife in the family plot at Trinity Cemetery in New York City.

FUN FACT

Seabury was nicknamed "the Bishop" due to his religious affiliations.

CEMETERY

Trinity Cemetery
550 West 155th Street
New York, NY 10032
Tel.: 212-368-1600
Hours: Mon.–Sun., 9 AM–4 PM

DIRECTIONS TO GRAVE

Enter the cemetery's westerly division at 770 Riverside Drive close to 153rd Street. You will then proceed to walk the length of the cemetery through labyrinthian paths going upward till you reach the end of the cemetery that borders Broadway. Just as you can see the street, you have reached the end of that path. From there, you are approximately in the middle area of the cemetery bordering Broadway. Follow the walkway west, straight about 50 feet, and you will see the Seabury plot on the left with Seabury's gravestone the third from the walkway. It has a cross on top.

OLIVER WENDELL HOLMES, JR.

U.S. Supreme Court Justice
Born: March 8, 1841
Died: March 6, 1935

Oliver Wendell Holmes, Jr. is considered one of the leading jurists and legal scholars to ever sit on the U.S. Supreme Court. He was well known for his pithy opinions, particularly on civil liberties and Amer-

Oliver Wendell Holmes, Jr.'s Grave

ican constitutional democracy. He was also a bona-fide civil war hero, being wounded a total of three times. Holmes set the record that still stands for being the oldest justice to retire from the Supreme Court at the age of 90.

Holmes was born in Boston to the prominent writer and physician Oliver Wendell Holmes, Sr. and abolitionist Amelia Lee Jackson. Holmes grew up in a rarified atmosphere of intellectual achievement and became lifelong friends with both Henry and William James, notable intellectuals of the day. Ralph Waldo Emerson was a close family friend. While at Harvard College, he wrote essays on philosophic themes and asked Emerson to read them.

While in his senior year at Harvard, Holmes enlisted at the outset of the Civil War in the Fourth Battalion of Infantry in the Massachusetts militia. Later, with his father's help, he secured a second lieutenant commission with the Twentieth Regiment of Massachusetts Volunteer Infantry. He saw extensive combat and suffered wounds at the Battle of Ball's Bluff, Antietam and Chancellorsville. In September 1863 while recovering at his family's home in Boston, he was promoted to colonel. Upon his recovery he was appointed aide-de-camp to General Horatio Wright and served with him during Grant's campaign down to Petersburg, Virginia. On July 17, 1864, he left the army, and enrolled at Harvard Law School later that year.

In 1866, he received his law degree and was admitted to the Massachusetts bar. He joined a small law firm and married a childhood friend, Fanny Bowditch Dixwell, in 1872. Their union lasted until her death in 1929. They were childless but adopted and raised an orphaned cousin, Dorothy Upham.

Holmes practiced admiralty and commercial law in Boston for 15 years, producing scholarly work during that period that culminated in the publication of *The Common Law* in 1881. Continuously in print since then, it remains controversial, with Holmes rejecting various kinds of formalism in law. His own view of the history of common law reads in part, "The life of the law has not been logic, it has been experience." Law as it evolved in modern societies was concerned with the material results of the defendant's actions. A judge's task was to decide which of the two parties before him bore the responsibility for the cost of the injury.

A vacancy occurred on the Massachusetts Supreme Judicial Court and Holmes was appointed to fill it by Governor John Davis Long on December 15, 1882. On August 2, 1899 he became the chief justice of said court. During his service on the court, he continued to develop and apply his views on the common law. He issued few constitutional opinions.

On December 2, 1902, President Theodore Roosevelt submitted the name of Holmes for nomination to the U.S. Supreme Court following the death of Associate Justice Horace Gray. Roosevelt reputedly admired Holmes' "Soldier's Faith" speech, which in part said, "That the joy of life is living, is to put out all one's powers as far as they will go; that the measure of power is obstacles overcome. . . ." He was confirmed and sworn into office on December 8, 1902.

Thus began a fruitful career on the nation's highest court. During his 29-plus years on the court, he ruled on cases spanning the whole range of federal law. His opinions ranged from copyright law, the law of contempt, the antitrust status of professional baseball, and even the oath required for citizenship. His tenure established him as one of the greatest judges in American history. Although he didn't dissent frequently—only 72 times versus 852 majority opinions—he earned the title of "the Great Dissenter" because his opinions were prescient and acquired so much authority.

Signature of Holmes, from the private collection of Vincent Gardino

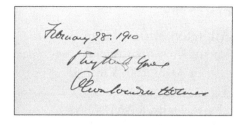

Urged on by his brethren on the court to retire due to his advanced age, he did so on January 12, 1932, just short of 91 years of age. He continued to be active. In 1933, a few days after his inauguration, the new President Franklin Roosevelt paid a courtesy call on Holmes. He found Holmes in his library reading Plato. FDR asked, "Why do you read Plato, Mr. Justice?" "To improve my mind, Mr. President," replied Holmes.

On February 23, 1935, a bitterly cold day, Justice Holmes went for a ride in his car with his secretary. The next day Holmes had a cold that rapidly developed into pneumonia. The nation held its breath waiting for reports as to his progress and was crushed when he died on March 5 in his home. His funeral at All Souls Church was followed by interment at Arlington. His committal ceremony was attended by the President and all his fellow justices. Three volleys of rifle shots were fired, one for each of his three Civil War wounds.

FUN FACTS

When his wife Fanny passed away in 1929, Holmes initially did not want a funeral, citing Fanny's wishes. Chief Justice William Howard Taft was fond of Fanny and would not hear of it, and he convinced Holmes to have one. Having yielded on this, Holmes let Taft take charge and, more importantly, Taft secured a burial plot for Holmes and Fanny in Arlington National Cemetery. Holmes had been too shy to ask the Secretary of War for this favor.

Upon his death, Holmes left his entire estate to the U.S. government. He had earlier said, "Taxes are what we pay for civilized society." His personal effects included his Civil War officer's uniform stained with his blood and "torn with shot" as well as the Minie balls, a type of bullet, that had wounded him three times in separate battles.

CEMETERY

Arlington National Cemetery
End of Memorial Avenue, which extends from the Memorial Bridge in Arlington, Virginia
Tel.: 877-907-8585
Hours: Daily, 8 AM–5 PM

DIRECTIONS TO GRAVE

From the Visitor's Center go straight to Schley Drive. Turn left on Schley to Roosevelt Drive. Follow to Weeks Road, where you turn right. Bear right on the paved walkway. Turn right on Section 5, which is where many Supreme Court justices are buried. Holmes' plot is located on a downward slope and visually difficult to find. It is near the road on the right of the section. To best spot it, look for the name "HOLMES" on one side of his gravestone.

Afterword

And now we have come to the end of our tour through American history with the Gardino brothers by visiting the graves of fifty-one Americans in all parts of the United States. These thirty-nine men and twelve women made careers in many different fields of endeavor. They were soldiers, judges, actors, singers, athletes, political leaders, clergymen. But they all had one thing in common—they used their talents to help others and to build a better society. To stand at their graves with Robert and Vince Gardino is to take inspiration from their lives.

The "fun facts" the Gardinos furnish about each of them help flesh out their humanity and their personalities. We learn that Clare Booth Luce coined the phrase "No good deed goes unpunished." But Vince Lombardi, we discover, did not say that "Winning isn't everything, it's the only thing," though his life certainly exemplified his competitive nature. However, Bob Hope did say, when he was asked on his deathbed where he would like to be buried, "Surprise me." Other surprises leaven the stories of the lives and deaths of the heroes and heroines in this book, which offers a unique way to learn our history.

JAMES McPHERSON
Pulitzer Prize–winning author
of *Battle Cry of Freedom*

Acknowledgments

The authors wish to acknowledge the following for their assistance and encouragement in helping us put together this second *Grave Trippers* volume. Those people who provided us with photographs are mentioned below and cited in the photo captions. All photos not credited were taken by the authors.

To our publisher, Edward Jutkowitz, for whom we will be forever grateful for allowing us to publish this second volume of *Grave Trippers* and for honoring us from the outset by allowing us to bring our hobby to a wider audience. Thank you for the confidence you have expressed in us.

To our copyeditor, Jennifer French, who has brought our writing in this book to a higher level with her expertise in cogent writing.

To Jerilyn DiCarlo for yet another great cover design for this volume of *Grave Trippers*.

To Robert and Patrice Martin, who transcend the meaning of friendship and always have their door open for us. They are family to us.

For Susan and Don Lukenbill, Grave Trippers for many years and cherished friends who assisted with photos and directions in Los Angeles.

For John and Melissa Capuano, who, in a diabolical moment, conceived the idea of *Grave Trippers* and who will always be an integral part of our lives (whether they want to be or not!).

For Monsignor Michael Crimmins, Father John Duffel, Father Douglas Crawford and Monsignor Robert Ritchie, who provide us with much needed spiritual guidance.

ACKNOWLEDGMENTS

For John Papa, superintendent at St. John's Episcopal Cemetery in Laurel Hollow, New York, for his expert guidance in navigating its grounds.

For Gino and Pinuccia Guasti, who always go out of their way to make us feel at home and welcome in Italy. Super amici!

For Gloria and Giuseppe Baldino, whose interest in history, literature and good cooking inspire us in Italia.

For Maan Wiccha, who always has our backs at the New York Athletic Club.

For Dominic Bruzzese, a true friend, who is always checking up on us.

For Dr. Justin Rashbaum and Irene Macaraeg, who both keep our teeth pearly white for book presentations.

For Walter Herbst, a fellow history buff, for his suggestions in the writing of this book.

For Charlie Salomon, whom we consider more like a member of the family than a friend.

For Vilma Lonsdale, our treasured cousin, who always brings a smile to our faces whenever we speak to one another. Viva Settu!

For Justin Lawrence, who is always the perfect gentleman to us—and always there with a smile for us at the NYAC!

For Florin and Dorina Isopescu, the reliable and trustworthy caretakers of our home in Italy. They are the best!

For Sheila and Tim Finnerty, who have been steadfast in their support of us and our endeavors.

For Roma Torre for her assistance in writing the biography on her mom.

For Ken and Kris Donovan, first and foremost great friends, and secondly, big *Grave Tripper* fans from the very beginning—their support has been always there and been appreciated!

For Carlo and Silvana Lovisolo of our favorite Italian restaurant, La Violetta. Unfortunately, it is not in New York, but is in Calamandrana, Italy. We try and get as much of their delicious home cooking as possible, inspired by Carlo's late mother Maria, each year on our visit, where they always make us feel as if we were in our own home.

For Giancarlo, Piera, Claudio and Maria Rupati. We so much miss their L'Angelo Café in Nizza Monferrato, an institution in Piedmont,

Italy. What we cherish most of all is their continued friendship, which never waivers.

For our close friends Antonella and Francesco Garino, who always bring us good cheer when they see us.

For MaryLou Falcone, cherished neighbor and friend, for all her help, insight and support.

For Veronique and Victor Callegari, steadfast neighbors and cherished friends.

For Sophia and Chang Luo, who have followed our adventures with such enthusiasm and shared our mutual love of chocolate cake!

For our devoted and cherished brother-in-law and sister-in-law, Frank and Judy Kim—they are truly members of our family.

For David Asman and his wife, M.C., for their unwavering spiritual support of all our endeavors.

Last, but not least, the Gaja family: Angelo, Lucia, Gaia, Rosanna and Giovanni. They have been enthusiastic supporters of our grave tripping activities during a friendship that has spanned nearly 30 years.

All royalties from our Grave Tripper series go to the Jinny Kim Gardino Memorial Scholarship Fund at Stephens College. If you care to make an individual donation, you may send it to:

>Shannon Walls
>Stephens College
>1200 East Broadway
>Campus Box 2035
>Columbia, MO 65215

Make your donation payable to Stephens College and note that it is for the Jinny Kim Gardino Scholarship Fund. Thank you!